Linear Elastic Waves

Wave propagation and scattering are among the most fundamental processes that we use to comprehend the world around us. Whereas these processes are often very complex, one way to begin to understand them is to study wave propagation in the linear approximation. This is a book describing such propagation using, as a context, the equations of elasticity. Two unifying themes are used. The first is that an understanding of plane wave interactions is fundamental to understanding more complex wave interactions. The second is that waves are best understood in an asymptotic approximation where they are free of the complications of their excitation and are governed primarily by their propagation environments. The topics covered include reflection, refraction, propagation of interfacial waves, integral representations, radiation and diffraction, and propagation in closed and open waveguides.

Linear Elastic Waves is an advanced level textbook directed at applied mathematicians, seismologists, and engineers.

John G. Harris received an undergraduate degree (honors) in 1971 from McGill University, where he studied electrical engineering and gained a lasting interest in wave propagation. In 1979, he received a doctoral degree in applied mathematics from Northwestern University for a thesis on elastic-wave diffraction problems. Following this, he joined the Theoretical and Applied Mechanics Department at the University of Illinois (Urbana-Champaign), where he teaches dynamics and mathematical methods and carries out research on elastic-wave problems at microwave frequencies. He has used asymptotic methods of analysis to explore diffraction and imaging, propagation and scattering of interfacial waves, and waveguiding.

Cambridge Texts in Applied Mathematics

Maximum and Minimum Principles
M. J. Sewell

Solitons
P. G. Drazin and R. S. Johnson

The Kinematics of Mixing
J. M. Ottino

Introduction to Numerical Linear Algebra and Optimisation
Philippe G. Ciarlet

Integral Equations
David Porter and David S. G. Stirling

Perturbation Methods
E. J. Hinch

The Thermomechanics of Plasticity and Fracture
Gerard A. Maugin

Boundary Integral and Singularity Methods for Linearized Viscous Flow
C. Pozrikidis

Nonlinear Wave Processes in Acoustics
K. Naugolnykh and L. Ostrovsky

Nonlinear Systems
P. G. Drazin

Stability, Instability and Chaos
Paul Glendinning

Applied Analysis of the Navier–Stokes Equations
C. R. Doering and J. D. Gibbon

Viscous Flow
H. Ockendon and J. R. Ockendon

Scaling, Self-Similarity, and Intermediate Asymptotics
G. I. Barenblatt

A First Course in the Numerical Analysis of Differential Equations
Arieh Iserles

Complex Variables: Introduction and Applications
Mark J. Ablowitz and Athanassios S. Fokas

Mathematical Models in the Applied Sciences
A. C. Fowler

Thinking About Ordinary Differential Equations
Robert E. O'Malley

A Modern Introduction to the Mathematical Theory of Water Waves
R. S. Johnson

Rarefied Gas Dynamics
Carlo Cercignani

Symmetry Methods for Differential Equations
Peter E. Hydon

High Speed Flow
C. J. Chapman

Wave Motion
J. Billingham and A. C. King

An Introduction to Magnetohydrodynamics
P. A. Davidson

Linear Elastic Waves
John G. Harris

Linear Elastic Waves

JOHN G. HARRIS
University of Illinois at Urbana-Champaign

CAMBRIDGE
UNIVERSITY PRESS

CAMBRIDGE
UNIVERSITY PRESS

32 Avenue of the Americas, New York NY 10013-2473, USA

Cambridge University Press is part of the University of Cambridge.

It furthers the University's mission by disseminating knowledge in the pursuit of education, learning and research at the highest international levels of excellence.

www.cambridge.org
Information on this title: www.cambridge.org/9780521643832

First published 2001

A catalogue record for this publication is available from the British Library

Library of Congress Cataloguing in Publication data
Harris, John G.
Linear elastic waves / John G. Harris.
p. cm. – (Cambridge texts in applied mathematics)
Includes bibliographical references and index.
ISBN 0-521-64368-6 – ISBN 0-521-64383-X (pb.)
1. Elastic waves. I. Title. II. Series.
QA935 .H227 2001
531′.1133 – dc21 00-065074

ISBN 978-0-521-64368-9 Hardback
ISBN 978-0-521-64383-2 Paperback

To Beatriz

Contents

Preface		*page* xiii
1	**Simple Wave Solutions**	**1**
1.1	Model Equations	1
	1.1.1 One-Dimensional Models	3
	1.1.2 Two-Dimensional Models	4
	1.1.3 Displacement Potentials	5
	1.1.4 Energy Relations	5
1.2	The Fourier and Laplace Transforms	6
1.3	A Wave Is Not a Vibration	11
1.4	Dispersive Propagation	13
	1.4.1 An Isolated Interaction	13
	1.4.2 Periodic Structures	15
References		18
2	**Kinematical Descriptions of Waves**	**20**
2.1	Time-Dependent Plane Waves	20
2.2	Time-Harmonic Plane Waves	22
2.3	Plane-Wave or Angular-Spectrum Representations	24
	2.3.1 A Gaussian Beam	24
	2.3.2 An Angular-Spectrum Representation of a Spherical Wave	26
	2.3.3 An Angular-Spectrum Representation of a Cylindrical Wave	28
2.4	Asymptotic Ray Expansion	28
	2.4.1 Compressional Wave	29
	2.4.2 Shear Wave	33

Appendix: Spherical and Cylindrical Waves 34
References 35

3 Reflection, Refraction, and Interfacial Waves 37
3.1 Reflection of a Compressional Plane Wave 37
3.1.1 Phase Matching 39
3.1.2 Reflection Coefficients 40
3.2 Reflection and Refraction 41
3.3 Critical Refraction and Interfacial Waves 44
3.4 The Rayleigh Wave 48
3.4.1 The Time-Harmonic Wave 49
3.4.2 Transient Wave 50
3.4.3 The Rayleigh Function 51
3.4.4 Branch Cuts 52
References 55

4 Green's Tensor and Integral Representations 56
4.1 Introduction 56
4.2 Reciprocity 57
4.3 Green's Tensor 58
4.3.1 Notes 61
4.4 Principle of Limiting Absorption 62
4.5 Integral Representation: A Source Problem 64
4.5.1 Notes 65
4.6 Integral Representation: A Scattering Problem 65
4.6.1 Notes 66
4.7 Uniqueness in an Unbounded Region 68
4.7.1 No Edges 68
4.7.2 Edge Conditions 69
4.7.3 An Inner Expansion 71
4.8 Scattering from an Elastic Inclusion in a Fluid 72
References 76

5 Radiation and Diffraction 77
5.1 Antiplane Radiation into a Half-Space 77
5.1.1 The Transforms 78
5.1.2 Inversion 79
5.2 Buried Harmonic Line of Compression I 82
5.3 Asymptotic Approximation of Integrals 86

5.3.1 Watson's Lemma 87
5.3.2 Method of Steepest Descents 90
5.3.3 Stationary Phase Approximation 94
5.4 Buried Harmonic Line of Compression II 96
5.4.1 The Complex Plane 96
5.4.2 The Scattered Compressional Wave 98
5.4.3 The Scattered Shear Wave 99
5.5 Diffraction of an Antiplane Shear Wave at an Edge 101
5.5.1 Formulation 102
5.5.2 Wiener–Hopf Solution 104
5.5.3 Description of the Scattered Wavefield 108
5.6 Matched Asymptotic Expansion Study 112
Appendix: The Fresnel Integral 116
References 119

6 Guided Waves and Dispersion 121
6.1 Harmonic Waves in a Closed Waveguide 121
6.1.1 Partial Waves and the Transverse Resonance Principle 123
6.1.2 Dispersion Relation: A Closed Waveguide 124
6.2 Harmonic Waves in an Open Waveguide 128
6.2.1 Partial Wave Analysis 129
6.2.2 Dispersion Relation: An Open Waveguide 131
6.3 Excitation of a Closed Waveguide 134
6.3.1 Harmonic Excitation 134
6.3.2 Transient Excitation 135
6.4 Harmonically Excited Waves in an Open Waveguide 139
6.4.1 The Wavefield in the Layer 140
6.4.2 The Wavefield in the Half-Space 145
6.4.3 Leaky Waves 146
6.5 A Laterally Inhomogeneous, Closed Waveguide 146
6.6 Dispersion and Group Velocity 150
6.6.1 Causes of Dispersion 150
6.6.2 The Propagation of Information 151
6.6.3 The Propagation of Angular Frequencies 153
References 157

Index 159

Preface

A wave
builds up
perhaps it says its name, I don't understand,
mutters, humps its load
of movement and foam
and withdraws. Who
can I ask what it said to me?[1]

Wave propagation and scattering are among the most fundamental processes that we use to comprehend the world around us. While these processes are often very complex, one way to begin to understand them is to study linear wave propagation. This is a book describing such propagation.

I use the equations of linear elasticity to form a context for my description of wave propagation. However, the reader's knowledge of elasticity need not be very great, and experience with a related field theory, such as fluid mechanics or electromagnetic theory, is sufficient to understand what is written here. In many places I treat only the antiplane shear problem because I do not believe that the extra work needed to do the analogous inplane problem adds anything of significance to understanding the underlying wave processes. Nevertheless, where an inplane elastic problem introduces a unique feature, such as the presence of a nondispersive surface wave, that problem is treated.

This is also a book describing the parts of applied mathematics that describe the propagation and scattering of linear elastic waves. It assumes that the reader has a good background in calculus, differential equations, and complex analysis. By this I mean that the reader should have studied most of the topics in

[1] Neruda, Pablo, *Soliloquy in the Waves*, pp. 185–186. In *Isla Negra, a Notebook*, translated by A. Reid. New York: Noonday Press. 1981.

Courant and John, *Introduction to Calculus and Analysis*, Vols. 1 and 2 (1989) and in Boyce and DiPrima, *Elementary Differential Equations and Boundary Value Problems* (1992). Moreover, the reader should be able to look things up in Carrier, Krook, and Pearson, *Functions of a Complex Variable* (1983) or Ablowitz and Fokas *Complex Variables, Introduction and Applications* (1997) and Zauderer *Partial Differential Equations of Applied Mathematics* (1998) and not feel lost. None of the mathematical analyses exceed, in sophistication or difficulty, those found in Courant and John (1989). I work almost entirely in Cartesian coordinates so that no knowledge of special functions or the transforms associated with them is needed. Some previous experience with asymptotic analysis would be helpful, but is not essential. All the asymptotic analysis needed is explained in the book.

I have used two unifying themes throughout. The first is that an understanding of plane-wave interactions is fundamental to understanding more complex wave interactions. The second is that waves are best understood in an asymptotic approximation where they are free of the complications of their excitation and are governed primarily by their propagation environment. Therefore plane-wave spectral analyses and asymptotic approximations are the main techniques used to study the more complicated problems.

The selection of problems for the reader is small and directed at engaging him or her in the development of the subject. The problems are an integral part of the book and most should be attempted.

I have tried to avoid a menagerie of symbols. In general I use Cartesian tensors such as τ_{ij}, where the indices $i, j = 1, 2, 3$, or a boldface notation $\boldsymbol{\tau}$. The symbol ∂_i is used to represent the partial derivative with respect to the ith coordinate. Similarly, sometimes I use $d_x f$ to represent df/dx. Repeated indices are summed over 1, 2, 3 unless otherwise indicated. For problems engaging only two coordinates, subscripts using Greek letters such as $\alpha, \beta = 1, 2$ are used so that a vector component would be written as u_β and a partial derivative as ∂_α. When these subscripts are repeated they are summed over 1, 2. At times I use symbols such as c_L or c_T when there is need to distinguish between parameters that relate to compressional or shear disturbances, but when that distinction is not important I drop the subscript. Constants such as A are used over and over again and have no special meaning.

Professor Jan Achenbach was my research advisor. He taught me the subject. His influence is everywhere is this book. I thank him. I have been supported over the years primarily by the National Science Foundation, though recently also by the Air Force Office of Scientific Research. I am very grateful for their support. Elaine Wilson typed the early parts of this manuscript and Mike

Greenberg provided the illustrations. Don Carlson and Bill Phillips tolerated my complaints with humor; Eduardo Velasco prevented me from publishing a discussion of group velocity that was at best muddled. To everyone, thank you.

The book undoubtedly contains errors, for which I alone am responsible. I anticipate that few if any are serious.

Books Cited

Ablowitz, M. J. and Fokas, A. S. 1997. *Complex Variables, Introduction and Applications.* New York: Cambridge.

Boyce, W. E. and DiPrima, R. C. 1992. *Elementary Differential Equations and Boundary Value Problems,* 5th ed. New York: Wiley.

Carrier, G. F., Krook, M., and Pearson, C. E. 1983. *Functions of a Complex Variable.* Ithaca, NY: Hod Books.

Courant, R. and John, F. 1989. *Introduction to Calculus and Analysis,* Vols. 1 and 2. New York: Springer.

Zauderer, E. 1983. *Partial Differential Equations of Applied Mathematics.* New York: Wiley-Interscience.

Evanston and Urbana, Summer 2000.

1

Simple Wave Solutions

Synopsis

Chapter 1 provides the background, both the model equations and some of the mathematical transformations, needed to understand linear elastic waves. Only the basic equations are summarized, without derivation. Both Fourier and Laplace transforms and their inverses are introduced and important sign conventions settled. The Poisson summation formula is also introduced and used to distinguish between a propagating wave and a vibration of a bounded body.

A linear wave carries information at a particular velocity, the group velocity, which is characteristic of the propagation structure or environment. It is this transmitting of information that gives linear waves their special importance. In order to introduce this aspect of wave propagation, propagation in one-dimensional periodic structures is discussed. Such structures are dispersive and therefore transmit information at a speed different from the wavespeed of their individual components.

1.1 Model Equations

The equations of linear elasticity consist of (1) the conservation of linear and angular momentum, and (2) a constitutive relation relating force and deformation. In the linear approximation density ρ is constant. The conservation of mechanical energy follows from (1) and (2). The most important feature of the model is that the force exerted across a surface, oriented by the unit normal n_j, by one part of a material on the other is given by the traction $t_i = \tau_{ji} n_j$, where τ_{ji} is the stress tensor. The conservation of angular momentum makes the stress tensor symmetric; that is, $\tau_{ij} = \tau_{ji}$. The conservation of linear momentum, in differential form, is expressed as

$$\partial_k \tau_{ki} + \rho f_i = \rho \partial_t \partial_t u_i. \tag{1.1}$$

The term $\boldsymbol{f}$ is a force per unit mass.

Deformation is described by using a strain tensor,

$$\epsilon_{ij} = (\partial_i u_j + \partial_j u_i)/2, \tag{1.2}$$

where u_i is ith component of particle displacement. The symmetrical definition of the deformation ensures that no rigid-body rotations are included. However, the underlying dependence of the deformation is upon the $\nabla \boldsymbol{u}$. For a homogeneous, isotropic, linearly elastic solid, stress and strain are related by

$$\tau_{ij} = \lambda \epsilon_{kk} \delta_{ij} + 2\mu \epsilon_{ij}, \tag{1.3}$$

where λ and μ are Lamé's elastic constants. Substituting (1.2) in (1.3), followed by substituting the outcome into (1.1), gives one form of the equation of motion, namely

$$(\lambda + \mu)\partial_i \partial_k u_k + \mu \partial_j \partial_j u_i + \rho f_i = \rho \partial_t \partial_t u_i\,. \tag{1.4}$$

Written in vector notation, the equation becomes

$$(\lambda + \mu)\nabla(\nabla \cdot \boldsymbol{u}) + \mu \nabla^2 \boldsymbol{u} + \rho\, \boldsymbol{f} = \rho \partial_t \partial_t \boldsymbol{u}. \tag{1.5}$$

When the identity $\nabla^2 \boldsymbol{u} = \nabla(\nabla \cdot \boldsymbol{u}) - \nabla \wedge \nabla \wedge \boldsymbol{u}$ is used, the equation can also be written in the form

$$(\lambda + 2\mu)\nabla(\nabla \cdot \boldsymbol{u}) - \mu \nabla \wedge \nabla \wedge \boldsymbol{u} + \rho\, \boldsymbol{f} = \rho \partial_t \partial_t \boldsymbol{u}. \tag{1.6}$$

This last equation indicates that elastic waves have both dilitational $\nabla \cdot \boldsymbol{u}$ and rotational $\nabla \wedge \boldsymbol{u}$ deformations.

If $\partial \mathcal{R}$ is the boundary of a region $\mathcal{R}$ occupied by a solid, then commonly $\boldsymbol{t}$ and $\boldsymbol{u}$ are prescribed on $\partial \mathcal{R}$. When $\boldsymbol{t}$ is given over part of $\partial \mathcal{R}$ and $\boldsymbol{u}$ over another part, the boundary conditions are said to be mixed. One very common boundary condition is to ask that $\boldsymbol{t} = 0$ everywhere on $\partial \mathcal{R}$. This models the case in which a solid surface is adjacent to a gas of such small density and compressibility that it is almost a vacuum. When $\mathcal{R}$ is infinite in one or more dimensions, special conditions are imposed such that a disturbance decays to zero at infinity or radiates outward toward infinity from any sources contained within $\mathcal{R}$.

Another common situation is that in which $\partial\mathcal{R}_{12}$ is the boundary between two regions, 1 and 2, occupied by solids having different properties. Contact between solid bodies is quite complicated, but in many cases it is usual to assume that the traction and displacement, $\boldsymbol{t}$ and $\boldsymbol{u}$, are continuous. This models welded contact. One other simple continuity condition that commonly arises is that between a solid and an ideal fluid. Because the viscosity is ignored, the tangential component of $\boldsymbol{t}$ is set to zero, while the normal component of traction and the normal component of displacement are made continuous.

These are only models and are often inadequate. To briefly indicate some of the possible complications, consider two solid bodies pressed together. A (linear) wave incident on such a boundary would experience continuity of traction and displacement when the solids press together, but would experience a traction-free boundary condition when they pull apart (Comninou and Dundurs, 1977). This produces a complex nonlinear interaction.

The reader may consult Hudson (1980) for a succinct discussion of linear elasticity or Atkin and Fox (1980) for a somewhat more general view.

1.1.1 One-Dimensional Models

We assume that the various wavefield quantities depend only on the variables x_1 and t. For *longitudinal strain*, u_1 is finite, while u_2 and u_3 are assumed to be zero, so that (1.2) combined with (1.3) becomes

$$\tau_{11} = (\lambda + 2\mu)\partial_1 u_1, \qquad \tau_{22} = \tau_{33} = \lambda\partial_1 u_1, \tag{1.7}$$

and (1.1) becomes

$$(\lambda + 2\mu)\partial_1\partial_1 u_1 + \rho f_1 = \rho\partial_t\partial_t u_1. \tag{1.8}$$

For *longitudinal stress*, all the stress components except τ_{11} are assumed to be zero. Now (1.3) becomes

$$\tau_{11} = E\partial_1 u_1, \qquad E = \mu\frac{3\lambda + 2\mu}{\lambda + \mu}, \tag{1.9}$$

and

$$\partial_2 u_2 = \partial_3 u_3 = -\nu\partial_1 u_1, \qquad \nu = \frac{\lambda}{2(\lambda + \mu)}. \tag{1.10}$$

Now (1.1) becomes

$$E\partial_1\partial_1 u_1 + \rho f_1 = \rho\partial_t\partial_t u_1. \tag{1.11}$$

Note that (1.8) and (1.11) are essentially the same, though they have somewhat different physical meanings. The longitudinal stress model is useful for rods having a small cross section and a traction-free surface. Stress components that vanish at the surface are assumed to be negligible in the interior.

1.1.2 Two-Dimensional Models

Let us assume that the various wavefield quantities are independent of x_3. As a consequence, (1.1) breaks into two separate equations, namely

$$\partial_\beta \tau_{\beta 3} + \rho f_3 = \rho \partial_t \partial_t u_3, \tag{1.12}$$

$$\partial_\beta \tau_{\beta\alpha} + \rho f_\alpha = \rho \partial_t \partial_t u_\alpha. \tag{1.13}$$

Greek subscripts $\alpha, \beta = 1, 2$ are used to indicate that the independent spatial variables are x_1 and x_2. The case for which the only nonzero displacement component is $u_3(x_1, x_2, t)$, namely (1.12), is called *antiplane shear* motion, or SH motion for shear horizontal.

$$\tau_{3\beta} = \mu \partial_\beta u_3, \tag{1.14}$$

giving, from (1.12),

$$\mu \partial_\beta \partial_\beta u_3 + \rho f_3 = \rho \partial_t \partial_t u_3. \tag{1.15}$$

Note that this is a two-dimensional scalar equation, similar to (1.8) or (1.11).

The case for which $u_3 = 0$, while the other two displacement components are generally nonzero, (1.13), is called *inplane motion.* The initials P and SV are used to describe the two types of inplane motion, namely compressional and shear vertical, respectively. For this case (1.3) becomes

$$\tau_{\alpha\beta} = \lambda \partial_\gamma u_\gamma \delta_{\alpha\beta} + \mu(\partial_\alpha u_\beta + \partial_\beta u_\alpha), \tag{1.16}$$

and

$$\tau_{33} = \lambda \partial_\gamma u_\gamma. \tag{1.17}$$

The equation of motion remains (1.4); that is,

$$(\lambda + \mu)\partial_\alpha \partial_\beta u_\beta + \mu \partial_\beta \partial_\beta u_\alpha + \rho f_\alpha = \rho \partial_t \partial_t u_\alpha. \tag{1.18}$$

This last equation is a vector equation and contains two wave types, compressional and shear, whose character we explore shortly. It leads to problems of some complexity.

These two-dimensional equations are the principal models used. The scalar model, (1.14), allows us to solve complicated problems in detail without being overwhelmed by the size and length of the calculations needed, while the vector model, (1.18), allows us enough structure to indicate the complexity found in elastic-wave propagation.

1.1.3 Displacement Potentials

When (1.4)–(1.6) are used, a boundary condition, such as $\boldsymbol{t}=0$, is relatively easy to implement. However, in problems in which there are few boundary conditions, it is often easier to cast the equations of motion into a simpler form and allow the boundary condition to become more complicated. One way to do this is to use the Helmholtz theorem (Phillips, 1933; Gregory, 1996) to express the particle displacement $\boldsymbol{u}$ as the sum of a scalar φ and a vector potential $\boldsymbol{\psi}$; that is,

$$\boldsymbol{u} = \nabla\varphi + \nabla\wedge\boldsymbol{\psi}, \qquad \nabla\cdot\boldsymbol{\psi} = 0. \tag{1.19}$$

The second condition is needed because $\boldsymbol{u}$ has only three components (the particular condition selected is not the only possibility). Assume $\boldsymbol{f}=0$. Substituting these expressions into (1.6) gives

$$(\lambda+2\mu)\nabla\left[\nabla^2\varphi - \left(1/c_L^2\right)\partial_t\partial_t\varphi\right] + \mu\nabla\wedge\left[\nabla^2\boldsymbol{\psi} - \left(1/c_T^2\right)\partial_t\partial_t\boldsymbol{\psi}\right] = 0. \tag{1.20}$$

The equation can be satisfied if

$$\nabla^2\varphi = \left(1/c_L^2\right)\partial_t\partial_t\varphi, \qquad c_L^2 = (\lambda+2\mu)/\rho, \tag{1.21}$$

$$\nabla^2\boldsymbol{\psi} = \left(1/c_T^2\right)\partial_t\partial_t\boldsymbol{\psi}, \qquad c_T^2 = \mu/\rho. \tag{1.22}$$

The terms c_L and c_T are the compressional or longitudinal wavespeed, and shear or transverse wavespeed, respectively. The scalar potential describes a wave of compressional motion, which in the plane-wave case is longitudinal, while the vector potential describes a wave of shear motion, which in the plane-wave case is transverse. Knowing φ and $\boldsymbol{\psi}$, do we know $\boldsymbol{u}$ completely? Yes we do. Proofs of completeness, along with related references, are given in Achenbach (1973).

1.1.4 Energy Relations

The remaining conservation law of importance is the conservation of mechanical energy. Again assume $\boldsymbol{f}=0$. This law can be derived directly from

(1.1)–(1.3) by taking the dot product of $\partial_t \boldsymbol{u}$ with (1.1). This gives, initially,

$$\partial_j \tau_{ji} \partial_t u_i - \rho(\partial_t \partial_t u_i) \partial_t u_i = 0. \tag{1.23}$$

Forming the product $\tau_{kl}\epsilon_{kl}$, using (1.3), and making use of the decomposition $\partial_j u_i = \epsilon_{ji} + \omega_{ji}$, where $\omega_{ji} = (\partial_j u_i - \partial_i u_j)/2$, allows us to write (1.23) as

$$\tfrac{1}{2}\partial_t(\rho \partial_t u_i \partial_t u_i + \tau_{ki}\epsilon_{ki}) + \partial_k(-\tau_{ki}\partial_t u_i) = 0. \tag{1.24}$$

The first two terms become the time rates of change of

$$\mathcal{K} = \tfrac{1}{2}\rho \partial_t u_k \partial_t u_k, \qquad \mathcal{U} = \tfrac{1}{2}\tau_{ij}\epsilon_{ij}. \tag{1.25}$$

These are the kinetic and internal energy density, respectively. The remaining term is the divergence of the energy flux, $\mathcal{F}$, where $\mathcal{F}$ is given by

$$\mathcal{F}_j = -\tau_{ji}\partial_t u_i. \tag{1.26}$$

Then (1.24) can be written as

$$\partial \mathcal{E}/\partial t + \nabla \cdot \mathcal{F} = 0, \tag{1.27}$$

where $\mathcal{E} = \mathcal{K} + \mathcal{U}$ and is the energy density. This is the differential statement of the conservation of mechanical energy. To better understand that the energy flux or power density is given by (1.26), consider an arbitrary region $\mathcal{R}$, with surface $\partial\mathcal{R}$. Integrating (1.27) over $\mathcal{R}$ and using Gauss' theorem gives

$$\frac{d}{dt}\int_{\mathcal{R}} \mathcal{E}(\boldsymbol{x}, t)\, dV = -\int_{\partial\mathcal{R}} \mathcal{F} \cdot \hat{\boldsymbol{n}}\, dS. \tag{1.28}$$

Therefore, as the mechanical energy decreases within $\mathcal{R}$, it radiates outward across the surface $\partial\mathcal{R}$ at a rate $\mathcal{F} \cdot \hat{\boldsymbol{n}}$.

1.2 The Fourier and Laplace Transforms

All waves are transient in time. One useful representation of a transient waveform is its Fourier one. This representation imagines the transient signal decomposed into an infinite number of time-harmonic or frequency components. One important reason for the usefulness of this representation is that the transmitter, receiver, and the propagation structure usually respond differently to the different frequency components. The linearity of the problem ensures that we can work out the net propagation outcomes for each frequency component and then combine the outcomes to recreate the received signal.

The Fourier transform is defined as

$$\overline{u}(x, \omega) = \int_{-\infty}^{\infty} e^{i\omega t} u(x, t)\, dt. \tag{1.29}$$

The variable ω is complex. Its domain is such as to make the above integral convergent. Moreover, $\overline{u}$ is an analytic function within the domain of convergence, and once known, can be analytically continued beyond it.[1] The inverse transform is defined as

$$u(x, t) = \frac{1}{2\pi} \int_{-\infty}^{\infty} e^{-i\omega t} \overline{u}(x, \omega)\, d\omega. \tag{1.30}$$

Thus we have represented u as a sum of harmonic waves $e^{-i\omega t}\overline{u}(x, \omega)$. *Note that there is a specific sign convention for the exponential term that we shall adhere to throughout the book.*

A closely related transform is the Laplace one. It is usually used for initial-value problems so that we imagine that for $t < 0$, $u(x, t)$ is zero. This is not essential and its definition can be extended to include functions whose t-dependence extends to negative values. This transform is defined as

$$\overline{u}(x, p) = \int_{0}^{\infty} e^{-pt} u(x, t)\, dt. \tag{1.31}$$

As with ω, p is a complex variable and its domain is such as to make $\overline{u}(x, p)$ an analytic function of p. The domain is initially defined as $\Re(p) > 0$, but the function can be analytically continued beyond this. Note that $p = -i\omega$ so that, when $t \in [0, \infty)$, $\Im(\omega) > 0$ gives the initial domain of analyticity for $\overline{u}(x, \omega)$. The inverse transform is given by

$$u(x, t) = \frac{1}{2\pi i} \int_{\epsilon - i\infty}^{\epsilon + i\infty} e^{pt} \overline{u}(x, p)\, dp, \tag{1.32}$$

where $\epsilon \geq 0$. The expressions for the inverse transforms, (1.30) and (1.32), are misleading. In practice we define the inverse transforms on contours that are designed to capture the physical features of the solution. A large part of this book will deal with just how those contours are selected. But, for the present, we shall work with these definitions.

[1] Analytic functions defined by contour integrals, including the case in which the contour extends to infinity, are discussed in Titchmarsh (1939) in a general way and in more detail by Noble (1988).

Consider the case of longitudinal strain. Imagine that at $t = -\infty$ a disturbance started with zero amplitude. Taking the Fourier transform of (1.8) gives

$$(d^2\bar{u}_1/dx_1^2) + k^2\bar{u}_1 = 0, \tag{1.33}$$

where $k = \omega/c_L$ and c_L is the compressional wavespeed defined in (1.21). The parameter k is called the wavenumber. Here (1.33) has solutions of the form

$$\bar{u}_1(x_1, \omega) = A(\omega)e^{\pm ikx_1}. \tag{1.34}$$

If we had sought a time-harmonic solution of the form

$$u_1(x_1, t) = \bar{u}_1(x_1, \omega)e^{-i\omega t}, \tag{1.35}$$

we should have gotten the same answer except that $e^{-i\omega t}$ would be present. In other words, taking the Fourier transform of an equation over time or seeking solutions that are time harmonic are two slightly different ways of doing the same operation.

For (1.35), it is understood that the real part can always be taken to obtain a real disturbance. Much the same happens in using (1.30). In writing (1.30) we implicitly assumed that $u(x, t)$ was real. That being the case, $\bar{u}(x, \omega) = \bar{u}^*(x, -\omega)$, where the superscript asterisk to the right of the symbol indicates the complex conjugate. From this it follows that

$$u(x, t) = \frac{1}{\pi}\Re\int_0^\infty e^{-i\omega t}\bar{u}(x, \omega)\, d\omega. \tag{1.36}$$

The advantage of this formulation of the inverse transform is that we may proceed with all our calculations by using an implied $e^{-i\omega t}$ and assuming ω is positive. The importance of this will become apparent in subsequent chapters. Now (1.36) can be regarded as a generalization of the taking of the real part of a time-harmonic wave (1.35).

Problem 1.1 Transform Properties

Check that $\bar{u}(x, \omega) = \bar{u}^*(x, -\omega)$ and derive (1.36) from (1.30). The reader may want to consult a book on the Fourier integral such as that by Papoulis (1962).

When the plus sign is taken, (1.35) is a time-harmonic, plane wave propagating in the positive x_1 direction. *We assume that the wavenumber k is positive, unless otherwise stated.* The wave propagates in the positive x_1 direction because the term $(kx_1 - \omega t)$ remains constant, and hence u_1 remains constant, only if x_1 increases as t increases. The speed with which the wave propagates is c_L. The term ω is the angular frequency or $2\pi f$, where f is the frequency. That is, at a fixed position, $1/f$ is the length of a temporal oscillation. Similarly, k, the wavenumber, is $2\pi/\lambda$, where λ, the wavelength, is the length of a spatial oscillation.

Note that if we combine two of these waves, labeled u_1^+ and u_1^-, each going in opposite directions, namely

$$u_1^+ = Ae^{i(kx_1 - \omega t)}, \qquad u_1^- = Ae^{-i(kx_1 + \omega t)}, \tag{1.37}$$

we get

$$u_1 = Ae^{-i\omega t} 2\cos(kx_1). \tag{1.38}$$

Taking the real part gives

$$u_1 = 2|A|\cos(\omega t + \alpha)\cos(kx_1). \tag{1.39}$$

We have taken A as complex so that α is its argument. This disturbance does not propagate. At a fixed x_1 the disturbance simply oscillates in time, and at a fixed t it oscillates in x_1. The wave is said to stand or is called a standing wave.

Problem 1.2 Fourier Transform

Continue with the case of longitudinal strain and consider the following boundary, initial-value problem. Unlike the previous discussion in which the disturbance began, with zero amplitude, at $-\infty$, here we shall introduce a disturbance that starts up at $t = 0^+$. Consider an elastic half-space, occupying $x_1 \geq 0$, subjected to a nonzero traction at its surface. The problem is one dimensional, and it is invariant in the other two so that (1.8), the equation for longitudinal strain, is the equation of motion. At $x_1 = 0$ we impose the boundary condition $\tau_{11} = -P_0\, e^{-\eta t} H(t)$, where $H(t)$ is the Heaviside step function and P_0 is a constant. As $x_1 \to \infty$ we impose the condition that any wave propagate outward in the positive x_1 direction. Why? Moreover, we ask that, for $t < 0$, $u_1(x_1, t) = 0$ and $\partial_t u_1(x_1, t) = 0$. Note that, using integration by parts, the Fourier transform, indicated by F, of the second time derivative is

$$F[\partial_t \partial_t u_1] = -\omega^2 \overline{u}_1(x_1, \omega) + i\omega u_1(x_1, 0^-) - \partial_t u_1(x_1, 0^-). \tag{1.40}$$

In deriving this expression we have integrated from $t = 0^-$ to ∞ so as to include any discontinuous behavior at $t = 0$. Taking the Fourier transform of (1.8) gives

(1.33). Show that the inverse transform of the stress component τ_{11} is given by

$$\tau_{11}(x_1, t) = \frac{P_0}{2\pi i} \int_{-\infty}^{\infty} \frac{e^{i(kx_1 - \omega t)}}{\omega + i\eta} \, d\omega. \tag{1.41}$$

In the course of making this step you will need to chose between the solutions to the transformed equation, (1.33). Why is the solution leading to (1.41) selected? Note that, if the disturbance is to decay with time, η must be positive. Next show that

$$\tau_{11}(x_1, t) = -P_0 H(t - x_1/c_L) e^{-\eta(t - x_1/c_L)}. \tag{1.42}$$

Explain how the conditions for convergence of the integral, as its contour is closed at infinity, give rise to the Heaviside function. Note how the sign conventions for the transform pair, by affecting where the inverse transform converges, give a solution that is causal.

Problem 1.3 Laplace Transform

Solve *Problem 1.2* by using the Laplace transform over time. Why select the solution e^{-px_1/c_L}? How does this relate to the demand that waves be outgoing at ∞?

The solution of *Problem 1.2* suggests how we shall define the Fourier transform over the spatial variable x. Suppose we have taken the temporal transform obtaining $\overline{u}(x, \omega)$. Then its Fourier transform over x is defined as

$$^*\overline{u}(k, \omega) = \int_{-\infty}^{\infty} e^{-ikx} u(x, t) \, dx, \tag{1.43}$$

and its inverse transform is

$$\overline{u}(x, \omega) = \frac{1}{2\pi} \int_{-\infty}^{\infty} e^{ikx} \, {}^*\overline{u}(k, \omega) \, dk. \tag{1.44}$$

Again note the sign conventions for the transform pair. *This will remain the convention throughout the book.* Moreover, note that

$$u(x, t) = \frac{1}{4\pi^2} \int_{-\infty}^{\infty} \int_{-\infty}^{\infty} e^{i(kx - \omega t)} \, {}^*\overline{u}(k, \omega) \, d\omega \, dk. \tag{1.45}$$

This shows that quite arbitrary disturbances can be decomposed into a sum of time-harmonic, plane waves and *thereby indicates that the study of such waves is very central to the understanding of linear waves.*

1.3 A Wave Is Not a Vibration

A continuous body vibrates when a system of standing waves is established within it. Vibration and wave propagation can be explicitly linked by means of the Poisson summation formula. This formula might better be termed a transform and is quite useful, especially for asymptotically approximating sums.

Proposition 1.1. *Consider a function $f(t)$ and its Fourier transform $\overline{f}(\omega)$. Restrictions on $f(t)$ are given below. The Poisson summation formula states that*

$$\sum_{m=-\infty}^{\infty} f(m\lambda) = \frac{1}{|\lambda|} \sum_{k=-\infty}^{\infty} \overline{f}\left(\frac{2\pi k}{\lambda}\right), \tag{1.46}$$

where λ is a parameter.

This formula relates a series to one comprising its transformed terms. If we want to approximate the left-hand side of (1.46) for λ that is small, then knowing the Fourier transform of each term enables us to use an approximation based on the fact that λ^{-1} is large. The left-hand side of (1.46) may converge only slowly for a small λ.

Proof.[2] This proof follows that of de Bruijn (1970). Consider the function $\varphi(x)$ given by

$$\varphi(x) = \sum_{m=-\infty}^{\infty} f[(m+x)\lambda], \tag{1.47}$$

where the sum converges uniformly on $x \in [0, 1]$. The function $\varphi(x)$ has period 1. We assume that $f(t)$ is such that $\varphi(x)$ has a Fourier series, $\varphi = \sum_{k=-\infty}^{\infty} c_k e^{ik2\pi x}$. Its kth Fourier coefficient equals

$$\begin{aligned}
\int_0^1 e^{-ik2\pi x}\varphi(x)\,dx &= \int_0^1 \sum_{m=-\infty}^{\infty} e^{-ik2\pi x} f[(m+x)\lambda]\,dx \\
&= \sum_{m=-\infty}^{\infty} \int_m^{(m+1)} e^{-ik2\pi x} f(x\lambda)\,dx \\
&= \frac{1}{|\lambda|} \int_{-\infty}^{\infty} e^{-ik2\pi(x/\lambda)} f(x)\,dx.
\end{aligned} \tag{1.48}$$

[2] A minimum amount of analysis is used, both here and elsewhere, and no attempt at rigorous proofs is made. The conclusions are usually valid under more general conditions than those cited.

Note that the integral $\int_m^{(m+t)} e^{-ikx} f(x\lambda)dx \to 0$ as $m \to \pm\infty$, uniformly in $x \in [0, 1]$, as the sum (1.47) converges uniformly. □

These conditions are more restrictive than needed. Lighthill (1978), among others, indicates that the Poisson summation formula holds for a far more general class of functions than assumed here.

Consider the impulsive excitation of a rod of finite length 1. The governing equation is (1.11). Assume $f_1 = 0$, set c_b and $\rho = 1$ ($c_b^2 = E/\rho$), and assume that for $t < 0$, $u_1(x_1, t) = \partial_t u_1(x_1, t) = 0$. The boundary conditions are

$$\tau_{11}(0, t) = -T\delta(t), \qquad \tau_{11}(1, t) = 0. \tag{1.49}$$

By using a Fourier transform over t and solving the boundary-value problem in x_1, we get[3]

$$\tau_{11} = iTH(t) \sum_{n=-\infty}^{\infty} e^{-in\pi t} \frac{\sin[n\pi(1 - x_1)]}{\cos(n\pi)}. \tag{1.50}$$

Thus the rod is filled with standing waves. One usually considers a solution in this form as a vibration. This is a very useful way to express the answer, assuming the pulse has reverberated within the rod for a time long with respect to that needed for one echo from the far end to return to $x = 0$. But the individual interactions with the ends have been obscured. To find these interactions we apply the Poisson summation formula to (1.50). Break up the $\sin[n\pi(1 - x_1)]$ term into exponential ones and apply (1.46) to each term. The crucial intermediate step is the following, where we have taken one of these terms.

$$\begin{aligned}\sum_{m=-\infty}^{\infty} \frac{1}{\cos m\pi} e^{-im\pi[t-(1-x_1)]} &= (\pi|t + x_1 - 2|)^{-1} \\ &\quad \times \sum_{n=-\infty}^{\infty} \int_{-\infty}^{\infty} e^{-i\omega\left(1 - \frac{2n}{|t+x_1-2|}\right)} d\omega \\ &= 2 \sum_{n=-\infty}^{\infty} \delta(|t + x_1 - 2| - 2n). \end{aligned} \tag{1.51}$$

The outcome to our calculation is

$$\tau_{11} = T\sum_{k=1}^{\infty} \delta(t + x_1 - 2k) - T\sum_{k=0}^{\infty} \delta(t - x_1 - 2k). \tag{1.52}$$

[3] The reader should check that this is the solution.

This is a wave representation. It is very useful for times of the order of the echo time. For example, if $t \in (1, 2)$, then $\tau_{11} = T\delta[(t-1)+(x_1-1)]$. We have thus isolated the pulse returning from its first reflection at the end $x_1 = 1$. However, the representation is not very useful for t that is large because it becomes tedious to work out exactly at what reflection you are tracking the pulse. Moreover, the representation (1.52) would have been awkward to work with if, instead of delta-function pulses, we had had pulses of sufficient length that they overlapped one another. Nevertheless, the representation captures quite accurately the physical phenomena of a pulse bouncing back and forth in a rod struck impulsively at one end.

A vibration therefore is defined and confined by its environment. It is the outcome of waves reverberating in a bounded body. A period of time, sometimes a long one, is needed for the environment to settle into a steady oscillatory motion. In contrast, a wave is a disturbance that propagates freely outward, returning to its source perhaps only once, experiencing only a finite number of interactions. Understanding how an individual wave interacts with its environment and tracking it through each of its interactions constitute the principal problems of wave propagation. Moreover, while one works frequently with time-harmonic propagating waves, one usually assumes that at some stage a Fourier synthesis will be carried out, mapping the unending oscillatory motion into a disturbance confined both temporally and spatially.

1.4 Dispersive Propagation

1.4.1 An Isolated Interaction

A basic interaction of a wave with its environment is scattering from a discontinuity. Continue to consider waves in a rod by using the longitudinal stress approximation. Consider time-harmonic disturbances of the form

$$u_1 = \overline{u}_1(x_1)e^{-i\omega t}, \qquad \tau_{11} = \rho c_b^2 \partial_1 u_1. \tag{1.53}$$

We shall not indicate the possible dependence on ω of $\overline{u}_1$ unless this is needed. The equation of motion (1.11) becomes

$$d^2\overline{u}_1/dx_1^2 + k^2\overline{u}_1 = 0, \qquad k = \omega/c_b. \tag{1.54}$$

Assume there is a region of inhomogeneity within $x_1 \in (-L, L)$. Incident on this inhomogeneity is the wavefield

$$\overline{u}_1^i(x_1) = \begin{cases} A_1 e^{ikx_1}, & x_1 < -L, \\ A_2 e^{-ikx_1}, & x_1 > L, \end{cases} \tag{1.55}$$

where we have allowed waves to be incident from both directions. Upon striking the inhomogeneity, the scattered wavefield

$$\overline{u}_1^s(x_1) = \begin{cases} B_1 e^{-ikx_1}, & x_1 < -L, \\ B_2 e^{ikx_1}, & x_1 > L \end{cases} \tag{1.56}$$

is excited. Note that the scattered waves have been constructed so that they are outgoing from the scatterer. The linearity of the problem suggests that we can write the scattered amplitudes in terms of the incident ones as

$$\begin{bmatrix} B_1 \\ B_2 \end{bmatrix} = \begin{bmatrix} S_{11} & S_{12} \\ S_{21} & S_{22} \end{bmatrix} \begin{bmatrix} A_1 \\ A_2 \end{bmatrix}, \tag{1.57}$$

or, more compactly, as

$$\boldsymbol{B} = \boldsymbol{S}\boldsymbol{A}. \tag{1.58}$$

The matrix $\boldsymbol{S}$ is called a scattering matrix and characterizes the inhomogeneity.

We next consider a specific example. Imagine that the rod has a cross-sectional area 1 and that the inhomogeneity is a point mass M, at $x_1 = 0$. The left-hand figure in Fig. 1.1 indicates the geometry. The conditions across the discontinuity are

$$u_1(0^-, t) = u_1(0^+, t), \qquad M\partial_t\partial_t u_1 = -\tau_{11}(0^-, t) + \tau_{11}(0^+, t). \tag{1.59}$$

That is, the rod does not break, but the acceleration of the mass causes the traction acting on the cross section to be discontinuous. Setting $\tan\theta = kM/2\rho$, with $\theta \in (0, \pi/2)$, we calculate the matrix $\boldsymbol{S}$ to be

$$\boldsymbol{S} = \begin{bmatrix} \sin\theta\, e^{i(\theta+\pi/2)} & \cos\theta\, e^{i\theta} \\ \cos\theta\, e^{i\theta} & \sin\theta\, e^{i(\theta+\pi/2)} \end{bmatrix}. \tag{1.60}$$

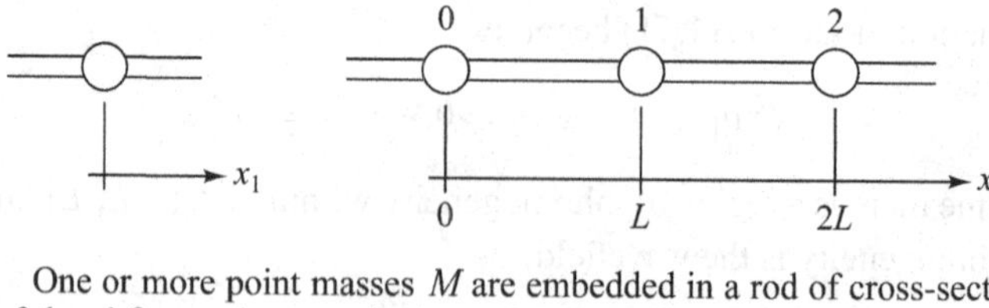

Fig. 1.1. One or more point masses M are embedded in a rod of cross-sectional area 1. The left-hand figure shows a single, embedded, point mass. The right-hand figure shows an endless periodic arrangement of embedded point masses, each separated by a distance L. The masses are labeled 0, 1, 2, ..., with the zeroth mass at $x_1 = 0$.

It is also of interest to relate the wave amplitudes on the right to those on the left. This matrix T, called the transmission matrix, gives $\boldsymbol{R} = \boldsymbol{T}\boldsymbol{L}$, where $\boldsymbol{L} = [A_1, B_1]^T$ and $\boldsymbol{R} = [B_2, A_2]^T$. The matrix $\boldsymbol{T}$ is readily calculated from $\boldsymbol{S}$ and is given by

$$T = \begin{bmatrix} 1 + i\tan\theta & i\tan\theta \\ -i\tan\theta & 1 - i\tan\theta \end{bmatrix}. \tag{1.61}$$

Note that the amplitudes A_1 and B_2 are those of right-propagating waves, while B_1 and A_2 are those of left-propagating ones.

Problem 1.4 Scattering From a Layer

Imagine a layer of width d and properties $(\bar{\rho}, \bar{c}_L)$ embedded in a material with properties (ρ, c_L). Let the x_1 axis be perpendicular to the face of the layer and place the origin at the left face, so that the layer occupies $x_1 \in (0, d)$. A wave of longitudinal strain Ae^{ikx_1} is incident from $x_1 < 0$. Calculate the reflection coefficient B/A and the transmission coefficient C/A, where the reflected wave is Be^{-ikx_1} and the transmitted one is $Ce^{ik(x_1-d)}$. Treat the waves in the layer as a sum of a single right and left propagating wave, each with different amplitudes, and imagine that the amplitudes contain the effects of all the multiple reflections from each side of the layer. Similarly B and C contain the effects of all the reflections and transmissions from the layer into the outer material. The wavenumber $k = \omega/c_L$. For what values of d is the reflection minimized? Can you identify any of the components of the $\boldsymbol{S}$ matrix for the layer in terms of these coefficients?

1.4.2 Periodic Structures

One of the more interesting aspects of wave-bearing structures is that they often contain several length scales. Propagation in such a structure often can only take place if the angular frequency ω is linked to the wavenumber – a term we must define a bit more carefully here – in a nonlinear way. To consider this possibility we use the matrix T, (1.61), to analyze propagation in a periodic structure. We imagine an infinite rod, cross-sectional area 1, in which equal point masses, M, are periodically embedded. The right-hand figure of Fig. 1.1 indicates the geometry and the how the masses are labeled. One such mass has the nominal position $x_1 = 0$ and is labeled $n = 0$. Each mass is separated from its neighbors by a distance L. A cell of length L is thereby formed and is labeled n if the nth mass occupies its left end. There are thus two length scales, the wavelength

$\lambda = 2\pi/k$ and the cell length L. In this problem we do not concern ourselves with how the waves are excited, but only with the simpler question, What waves does this structure support? Consider the zeroth cell, where $x_1 \in (0, L)$. Within that cell the solution to (1.54) is

$$\bar{u}_1(x_1) = R_0 e^{ikx_1} + L_0 e^{-ikx_1}. \tag{1.62}$$

At $x_1 = L^-$ the wavefield is $[R_0 e^{ikL}, L_0 e^{-ikL}]^T$. This can be written as $\boldsymbol{L}_1 = \boldsymbol{P}\boldsymbol{R}_0$, with $\boldsymbol{R}_0 = [R_0, L_0]^T$. The matrix $\boldsymbol{P}$ is called the propagator or the propagation matrix and is given by

$$\boldsymbol{P} = \begin{bmatrix} e^{ikL} & 0 \\ 0 & e^{-ikL} \end{bmatrix}. \tag{1.63}$$

At $x_1 = L^+$, within the 1th cell, the wavefield amplitudes $\boldsymbol{R}_1 = [R_1, L_1]^T$ are

$$\boldsymbol{R}_1 = \boldsymbol{T}\boldsymbol{P}\boldsymbol{R}_0. \tag{1.64}$$

This relation is readily generalized. If $\boldsymbol{R}_n = [R_n, L_n]^T$, then

$$\boldsymbol{R}_{n+1} = \boldsymbol{T}\boldsymbol{P}\boldsymbol{R}_n. \tag{1.65}$$

The central feature of the propagation structure is that it has translational symmetry. The central feature of the disturbance we seek is that its phase changes from cell to cell in a way that represents propagation. Specifically consider propagation to the right. To capture these two features, the wavefield at a point within the $(n+1)$th cell can differ from that at a point within the nth cell, where the two points are separated by a distance L, by at most a multiplicative phase factor. This kinematic constraint is expressed by the relation

$$\boldsymbol{R}_{n+1} = e^{i\kappa L} \boldsymbol{R}_n, \tag{1.66}$$

where κ is unknown. κ is positive, if real, and such as to cause decay, if complex. Combining (1.65) and (1.66) gives a 2×2 system of algebraic equations that has a nontrivial solution if and only if

$$\det(\boldsymbol{T}\boldsymbol{P} - e^{i\kappa L}\boldsymbol{I}) = 0. \tag{1.67}$$

Recalling our previous definition of $\tan\theta = kM/2\rho$, we can write this equation compactly as

$$\cos\kappa L = \cos(kL + \theta)/\cos\theta. \tag{1.68}$$

This is a nonlinear relationship between the angular frequency $\omega = c_b k$ and the effective wavenumber κ, though it may not be apparent, as yet, that κ (and not k) is the wavenumber of interest.

Note that if $\kappa L \in [-\pi, \pi]$ is a solution, then $\kappa L \pm 2n\pi$, for $n = 1, 2, \ldots$ is also a solution. Accordingly, we need only consider $\kappa L \in [-\pi, \pi]$. The term κL is real provided $|\cos \kappa L| \leq 1$. Therefore the boundaries between real and complex κL are given by

$$\cos(kL + \theta)/\cos\theta = \pm 1. \tag{1.69}$$

When $+1$ is taken, the solutions are $\sin(kL/2) = 0$ or $\tan(kL/2) = -\tan(\theta)$. That is, $kL = 2n\pi$ or $kL + 2\theta = 2m\pi$, where n and m are integers. When -1 is taken, the solutions are $\cos(kL/2) = 0$ or $\cot(kL/2) = \tan(\theta)$. That is, $kL = (2n-1)\pi$ or $kL + 2\theta = (2m-1)\pi$, where, again, n and m are integers. All these cases are covered by $kL = n\pi$ or $kL + 2\theta = m\pi$. For $kL \in [(n-1)\pi, (n\pi - 2\theta)]$, κL is real. These intervals are called passbands. Elsewhere κL is complex, causing the disturbance to decay as it propagates, and the intervals are called stopbands. At the lower boundary of a passband, L is an integer number of half-wavelengths. If T were real, then all the reflected waves add constructively and little or nothing is transmitted. The actual situation is complicated by the complex T, but the constructive interference of the reflected waves is the basic physical mechanism giving rise to the stopbands. This phenomenon is referred to as *Bragg scattering.*

Consider the interval $x_1 \in (nL, (n+1)L)$. Then

$$\begin{aligned}\bar{u}_1(x_1) &= R_n e^{ik(x_1-nL)} + L_n e^{-ik(x_1-nL)} \\ &= e^{i\kappa L}\left[R_{n-1} e^{ik(x_1-nL)} + L_{n-1} e^{-ik(x_1-nL)}\right] \\ &= e^{i\kappa L}\bar{u}_1(x_1 - L)\end{aligned} \tag{1.70}$$

This equation is a restatement of (1.66). Further, it indicates that $\bar{u}_1(x_1)$ must satisfy the functional equation $\bar{u}_1(x_1 + L) = e^{i\kappa L}\bar{u}_1(x_1)$ if the kinematic constraint is to be enforced. Within each cell there are nominally two waves, as indicated in (1.70), that we call partial waves. However, we seek a solution for the wave globally propagating to the right along the structure, as distinguished from the right- and left-propagating partial waves in each cell. With this in mind, the solution to the functional equation is

$$\bar{u}_1(x_1) = e^{i\kappa x_1}\varphi(x_1), \tag{1.71}$$

where $\varphi(x_1 + L) = \varphi(x_1)$[4]. That is, $\varphi(x_1)$ is a periodic function and can be represented by a Fourier series, whose coefficients are c_n. Therefore, $\bar{u}_1(x_1)$ becomes

$$\bar{u}_1(x_1) = \sum_{-\infty}^{\infty} c_n e^{ix_1(\kappa - 2\pi n/L)}. \tag{1.72}$$

The time-harmonic wavefield $\bar{u}_1(x_1)e^{-i\omega t}$ is thus a consequence of an infinite number of space harmonics. Note that shifting κL by $\pm 2m\pi$ would not change this expression.

More importantly, it is clear that it is κ, through the term $e^{i(\kappa x_1 - \omega t)}$, that is the wavenumber. Equation (1.68) indicates that ω is a function of κ, or κ a function of ω. A relation such as this is called a dispersion relation. Writing κ as $\omega/c(\omega)$, we see that $\bar{u}_1(x_1, \omega)$ propagates at a different speed for each ω. If we excited the structure with a pulse, then the pulse would be composed of an infinite number of such components, as indicated by (1.36). Each component would then propagate at its own speed and the pulse would become dispersed. A pulse is information, whereas a sinusoid is not. Hence what we have inferred is that dispersion can cause the distortion of or loss of information from a signal. We shall explore this topic further in Chapter 6.

There are many fascinating aspects to propagation in periodic structures. The discussion here has followed parts of Levine (1978), and a reader seeking to learn more may wish to read this work further.

References

Achenbach, J.D. 1973. *Wave Propagation in Elastic Solids.* Amsterdam: North-Holland.

Atkin, R.J. and Fox, N. 1980. *An Introduction to the Theory of Elasticity.* London: Longman.

de Bruijn, N.G. 1970. *Asymptotic Methods in Analysis,* 3rd ed., pp. 52–56. Amsterdam: North-Holland.

Comninou, M. and Dundurs, J. 1977. Reflexion and refraction of elastic waves in the presence of separation. *Proc. R. Soc. Lond.,* A, **356**: 509–528.

Friedman, B. 1956. *Principles and Techniques of Applied Mathematics,* pp. 65–67. New York: Wiley.

Gregory, R.D. 1996. Helmholtz's theorem when the domain is infinite and when the field has singular points. *Quart. J. Mech. Appl. Math.* **49**:439–450.

Hudson, J.A. 1980. *The Excitation and Propagation of Elastic Waves.* Cambridge: University Press.

Levine, H. 1978. *Unidirectional Wave Motions,* pp. 273–308, 339–345, and elsewhere. Amsterdam: North-Holland.

[4] This is a partial statement of Floquet's theorem (Friedman, 1956).

Lighthill, M.J. 1978. *Fourier Analysis and Generalized Functions*, pp. 67–71. Cambridge: University Press.

Noble, B. 1988. *Methods Based on the Wiener-Hopf Technique*, pp. 11–27. New York: Chelsea.

Papoulis, A. 1962. *The Fourier Integral and its Applications*. New York: McGraw-Hill.

Phillips, H.B. 1933. *Vector Analysis*, pp. 182–196. New York: Wiley.

Titchmarsh, E.C. 1939. *The Theory of Functions*, 2nd ed., pp. 85–86. Oxford: Clarendon Press.

2

Kinematical Descriptions of Waves

Synopsis

A wave equation is an equation of motion, so that a study of its solutions is perhaps more than just the study of the kinematics. However, when one writes little about how a wavefield is excited and concentrates primarily upon the geometrical description of a wavefield's propagation, it is reasonable to label this as such. The present chapter describes the kinematics of waves in a propagation environment without boundaries. Descriptions of a plane wave are emphasized because they form a canonical kinematical object from which more complicated solutions can be constructed.

2.1 Time-Dependent Plane Waves

A time-dependent plane wave is one whose form is

$$\boldsymbol{u}(\boldsymbol{x}, t) = \hat{\boldsymbol{d}}\, u(t - \boldsymbol{s} \cdot \boldsymbol{x}), \tag{2.1}$$

where $\hat{\boldsymbol{d}}$ is a constant unit vector[1] called the polarization and $\boldsymbol{s}$ is a constant vector called the slowness. The time is t and the position vector is $\boldsymbol{x}$. The equation $\boldsymbol{s} \cdot \boldsymbol{x} = t - C$, where C is a constant, is a parametric representation of a family of planes, where each value of parameter t gives a different member of the family. The normal to the plane is $\boldsymbol{s}$. Figure 2.1 sketches this relationship. As t is incremented, then $\boldsymbol{x}$ must be incremented along $\boldsymbol{s}$ if the argument $t - \boldsymbol{s} \cdot \boldsymbol{x}$ is to equal the constant C. Accordingly, u retains a constant value over each plane defined by t provided this plane advances along the normal $\boldsymbol{s}$ as t increases. The wave's name then arises from this fact and the plane is called a wave front. Setting $\boldsymbol{s} = s\hat{\boldsymbol{p}}$ so that $\hat{\boldsymbol{p}}$ is the unit normal to each plane, it is then

[1] The circumflex indicates a unit vector.

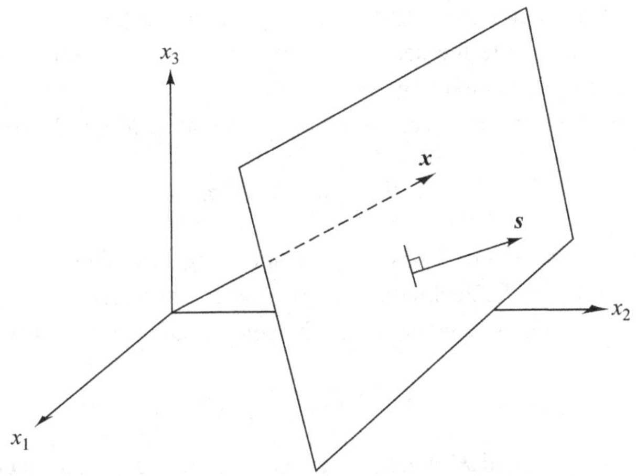

Fig. 2.1. A sketch of the plane wavefront $\boldsymbol{s} \cdot \boldsymbol{x} = t - C$, where t is a parameter identifying a particular plane. The position vector $\boldsymbol{x}$ identifies a point on the plane; the slowness vector $\boldsymbol{s}$ is normal to the plane and hence gives its orientation.

straightforward to find that

$$\hat{\boldsymbol{p}} \cdot d\boldsymbol{x}/dt = 1/s, \tag{2.2}$$

so that s^{-1} is the speed of advance of the plane along $\hat{\boldsymbol{p}}$.

Substituting (2.1) into the equation of motion, (1.6), gives

$$c_L^2 \boldsymbol{s}(\boldsymbol{s} \cdot \partial_t \partial_t \boldsymbol{u}) - c_T^2 \boldsymbol{s} \wedge \boldsymbol{s} \wedge \partial_t \partial_t \boldsymbol{u} = \partial_t \partial_t \boldsymbol{u}. \tag{2.3}$$

Because, for any vector $\boldsymbol{A}$, $\boldsymbol{s}(\boldsymbol{s} \cdot \boldsymbol{A}) - \boldsymbol{s} \wedge \boldsymbol{s} \wedge \boldsymbol{A} = s^2 \boldsymbol{A}$, (2.3) implies that either

$$s^2 = 1/c_L^2, \qquad \boldsymbol{s} \wedge \boldsymbol{u} = 0, \tag{2.4}$$

or

$$s^2 = 1/c_T^2 \qquad \boldsymbol{s} \cdot \boldsymbol{u} = 0. \tag{2.5}$$

Thus either

$$\boldsymbol{u} = \hat{\boldsymbol{d}} u(t - \hat{\boldsymbol{p}} \cdot \boldsymbol{x}/c_L), \qquad \hat{\boldsymbol{p}} \wedge \hat{\boldsymbol{d}} = 0, \tag{2.6}$$

or

$$\boldsymbol{u} = \hat{\boldsymbol{d}} u(t - \hat{\boldsymbol{p}} \cdot \boldsymbol{x}/c_T), \qquad \hat{\boldsymbol{p}} \cdot \hat{\boldsymbol{d}} = 0. \tag{2.7}$$

The first is a longitudinal wave with its polarization parallel to its direction of propagation, while the second is a transverse wave with its polarization perpendicular to this direction.

The flux of energy $\mathcal{F}$ is given by (1.26). Calculating $\mathcal{F}$ for (2.6) or (2.7) gives

$$\mathcal{F}_I = \rho c_I (\partial_t u)^2 \,\hat{\boldsymbol{p}}, \tag{2.8}$$

where $I = L$ or T and $\mathcal{F}_I$ is the corresponding flux. The energy density is $\mathcal{E} = \mathcal{K} + \mathcal{U}$, where $\mathcal{K}$, the kinetic energy, and $\mathcal{U}$, the internal energy, are given by (1.25). For these two plane waves the velocity of energy transport, $\boldsymbol{\mathcal{C}}$, is

$$\boldsymbol{\mathcal{C}}_I = \mathcal{F}_I/\mathcal{E}_I = c_I \hat{\boldsymbol{p}}, \tag{2.9}$$

where $I = L$ or T, and $\mathcal{F}_I$ and $\mathcal{E}_I$ are the corresponding flux and energy density.

2.2 Time-Harmonic Plane Waves

In one sense, time-harmonic plane waves are only a particular case of a general plane wave, but because we are working with the temporal transform, we have the additional flexibility of allowing $\hat{\boldsymbol{p}}$ to be complex. This has some interesting implications for the form of the corresponding time-dependent plane wave that we shall explore in Section 3.3. For the present, assume that the time dependence is $e^{-i\omega t}$ or that the wavefield has been transformed over time. The $e^{-i\omega t}$ is not given, unless explicitly needed, and ω is assumed real and positive, for the present.

A time-harmonic plane wave has the form[2]

$$\boldsymbol{u} = A\hat{\boldsymbol{d}} e^{ik\hat{\boldsymbol{p}}\cdot\boldsymbol{x}}, \tag{2.10}$$

where the amplitude $A = |A|e^{i\theta}$ and the wavenumber $k = \omega/c$. The combination $k\hat{\boldsymbol{p}} = \omega\boldsymbol{s}$ is the wavevector. To recover a real, time-harmonic wave, multiply by $e^{-i\omega t}$ and take the real part of the expression.

As indicated previously, we may take $\hat{\boldsymbol{p}} = \boldsymbol{p}_r + i\boldsymbol{p}_i$, with $\boldsymbol{p}_r$ and $\boldsymbol{p}_i$ both real. Because $\hat{\boldsymbol{p}} \cdot \hat{\boldsymbol{p}} = 1$, $\boldsymbol{p}_r \cdot \boldsymbol{p}_i = 0$. The real and imaginary components are perpendicular to one another. If $\boldsymbol{p}_i \neq 0$, the plane wave is an inhomogeneous or evanescent one, while if $\boldsymbol{p}_i = 0$, it is homogeneous. Writing (2.10) in detail gives

$$\boldsymbol{u} = A\hat{\boldsymbol{d}} e^{-k\boldsymbol{x}\cdot\boldsymbol{p}_i} e^{ik\boldsymbol{x}\cdot\boldsymbol{p}_r}. \tag{2.11}$$

[2] In the previous chapter we distinguished a time-harmonic wave from a time-dependent one by using an overbar. The overbar is no longer used unless explicitly needed.

An inhomogeneous plane wave propagates in the $\boldsymbol{p}_r$ direction, but decays in the $\boldsymbol{p}_i$ direction. That such waves arise in practice will soon become apparent. Substituting (2.10) into the equation of motion, (1.6), gives, just as it did in (2.3),

$$(c_L^2/c^2)\,\hat{\boldsymbol{p}}(\hat{\boldsymbol{p}}\cdot\hat{\boldsymbol{d}}) - (c_T^2/c^2)\,\hat{\boldsymbol{p}}\wedge\hat{\boldsymbol{p}}\wedge\hat{\boldsymbol{d}} = \hat{\boldsymbol{d}}. \tag{2.12}$$

Thus, either $\hat{\boldsymbol{p}}\wedge\hat{\boldsymbol{d}} = 0$ and $c = c_L$, or $\hat{\boldsymbol{p}}\cdot\hat{\boldsymbol{d}} = 0$ and $c = c_T$. Note that these statements imply that $\hat{\boldsymbol{d}}$ may also be complex.

The symbol $\hat{p}_i$ is the ith component of $\hat{\boldsymbol{p}}$. Assume that $\hat{p}_3 = 0$ and $\hat{p}_1$ is real. Then $\hat{p}_2$ may be real or imaginary. That is,

$$\hat{p}_2 = \begin{cases} \left(1 - \hat{p}_1^2\right)^{1/2}, & |\hat{p}_1| < 1, \\ i\left(\hat{p}_1^2 - 1\right)^{1/2}, & |\hat{p}_1| > 1. \end{cases} \tag{2.13}$$

In the first case the wave is homogeneous:

$$\boldsymbol{u} = A\hat{\boldsymbol{d}}e^{ik(x_1\hat{p}_1 + x_2\hat{p}_2)}. \tag{2.14}$$

In the second it is inhomogeneous:

$$\boldsymbol{u} = A\hat{\boldsymbol{d}}e^{-k\hat{\alpha}x_2}e^{ikx_1\hat{p}_1}. \tag{2.15}$$

Here $\alpha = -i\hat{p}_2$. The implications of allowing the vectors $\hat{\boldsymbol{p}}$ and $\hat{\boldsymbol{d}}$ to be complex lead to many interesting results, many of which are discussed by Boulanger and Hayes (1993).

For the remainder of the section, we shall assume that $\hat{\boldsymbol{p}}$ is real. Calculating the instantaneous energy flux gives

$$\mathcal{F}_I = \rho c_I \omega^2 \Re(iAe^{i\eta_I})\Re(iAe^{i\eta_I})\,\hat{\boldsymbol{p}}, \tag{2.16}$$

where $\eta_I = k_I(\boldsymbol{x}\cdot\hat{\boldsymbol{p}} - c_I t)$, and $I = L$ or T. The instantaneous value is of less interest than the average value over a period $T(= 2\pi/\omega)$. This average is defined as

$$\langle G\rangle = \frac{1}{T}\int_t^{t+T} G(\tau)\,d\tau, \tag{2.17}$$

where $G(t)$ is a *real* periodic function of t with period T. It can be shown by direct calculation that

$$\langle \Re(Ae^{i\eta})\Re(B^{i\eta})\rangle = \tfrac{1}{2}\Re(AB^*), \tag{2.18}$$

where A and B are the complex amplitudes and η has the form given previously.

Problem 2.1

Carry out the calculation leading to (2.18).

Applying (2.18) to (2.16) gives

$$\langle \mathcal{F}_I \rangle = \tfrac{1}{2}\rho c_I \omega^2 A A^* \hat{\boldsymbol{p}}. \tag{2.19}$$

Problem 2.2

Show that $\langle \mathcal{F}_I \rangle = c_I \langle \mathcal{E}_I \rangle \hat{\boldsymbol{p}}$ and $\mathcal{F}_I$ and $\mathcal{E}_I$ are the corresponding flux and energy density.

2.3 Plane-Wave or Angular-Spectrum Representations

Spherical and cylindrical waves can be constructed directly from solutions to the equations of motion in the corresponding coordinate systems. A summary of this approach is given in an Appendix. However, these waves can also be constructed from collections of homogeneous and inhomogeneous plane waves. These representations are called plane-wave or angular-spectrum representations. The advantage of these representations is that, upon constructing the outcome of an interaction of a plane wave with an obstacle, linearity allows us to use it to construct the outcome of an interaction with the same obstacle by a more general wavefield. Here, we construct an antiplane shear, Gaussian beam and a compressional spherical wave, and give the result for an antiplane shear, cylindrical wave. This last construction is taken up in *Problem 4.1*, because it is more easily motivated by starting from a radiation problem. The clarity of the plane-wave spectral technique is well demonstrated by Clemmow (1966) through the description and solution of several electromagnetic wave propagation and scattering problems.

2.3.1 A Gaussian Beam

Consider an antiplane shear wave whose equation of motion is (1.15). We shall continue to use u_3 for the particle displacement, but write c rather than the awkward c_T. When a time-harmonic disturbance is assumed, (1.15) gives, in a region where there are no sources of excitation,

$$\partial_\alpha \partial_\alpha u_3 + k^2 u_3 = 0. \tag{2.20}$$

We consider a signaling problem. This is a problem wherein the source is given as a function of time over an initial plane – in the present case it is given as a time-harmonic function along the x_2 axis – and the excited wavefield is allowed to propagate outward, usually in only one direction, to infinity. At $x_1 = 0$,

$$u_3(0, x_2) = Ae^{-(x_2/b)^2}, \tag{2.21}$$

where A is a complex amplitude and b is a parameter that measures the initial width of the wavefield. While it will not be clear from what we do here, this wavefield will remain Gaussian in cross section but spreads as it propagates outward. As $x_1 \to \infty$, we ask that the wave be outgoing (there are no sources at infinity).

When the spatial Fourier transform over x_2 is used, the equation of motion becomes

$$d^{2*}u_3/dx_1^2 + k_1^{2\,*}u_3 = 0, \tag{2.22}$$

where

$$k_1 = \left(k^2 - k_2^2\right)^{1/2}, \quad \Re(k_1) \geq 0, \quad \Im(k_1) \geq 0. \tag{2.23}$$

Defining the branches of the radical k_1 is very important, but for the present we do not need to know in detail where the branch cuts are placed. This will be discussed in Section 3.4.4. The solution to (2.22) that satisfies the outgoing condition is ${}^*u_3 = U(k_2)e^{ik_1x_1}$. Therefore

$$u_3(x_1, x_2) = \frac{1}{2\pi}\int_{-\infty}^{\infty} U(k_2)e^{i(k_1x_1+k_2x_2)}dk_2. \tag{2.24}$$

Note that we could simply have asserted that this was a solution to (2.20) and hence begun the discussion at this point.

To find U, we use (2.21) to write

$$\begin{aligned} U(k_2) &= A\int_{-\infty}^{\infty} e^{-(x_2/b)^2}e^{-ik_2x_2}dx_2, \\ &= \pi^{1/2}Abe^{-(k_2b/2)^2}. \end{aligned} \tag{2.25}$$

Therefore,

$$u_3(x_1, x_2) = \frac{bA}{2\pi^{1/2}}\int_{-\infty}^{\infty} e^{-(k_2b/2)^2}e^{i(k_1x_1+k_2x_2)}dk_2. \tag{2.26}$$

We have succeeded in taking a rather general wavefield and expressing it as an integral over a set of plane waves; hence the name plane-wave representation and the descriptive term plane-wave spectra. Note that k_1 takes both real and imaginary values as k_2 ranges over $(-\infty, \infty)$. That is, the integral is one over both homogeneous and inhomogeneous, time harmonic, plane waves.

2.3.2 An Angular-Spectrum Representation of a Spherical Wave

Consider a spherically symmetric compressional wave such that the particle displacement $\boldsymbol{u}$ is described by $\nabla\varphi$. The equation governing φ is (1.21). The only spatial dependence is upon $r = (x_1^2 + x_2^2 + x_3^2)^{1/2}$. Again we assume that the disturbance is time harmonic. For $r > 0$, (1.21) becomes

$$\frac{1}{r^2}\frac{\partial}{\partial r}\left(r^2\frac{\partial\varphi}{\partial r}\right) + k^2\varphi = 0, \tag{2.27}$$

where we have written c to replace c_L and $k = \omega/c$ to replace k_L. For $r \neq 0$ a solution of (2.27) is

$$\varphi = A(e^{ikr}/kr), \tag{2.28}$$

where A is an amplitude that may be complex. Clearly its wavefront is spherical. The energy flux is proportional to r^{-2} balancing the increase in surface area of the spherical wavefront as the wave propagates outward. In Section 4.3.1, *Note 2*, we suggest a possible source for such a spherical, compressional wave.

Rather than consider the wave excited by a compact source, we pose a signaling problem. Set $\rho = (x_1^2 + x_2^2)^{1/2}$. The potential φ satisfies (2.27), while, at $x_3 = 0$,

$$\varphi = A(e^{ik\rho}/k\rho). \tag{2.29}$$

As $x_3 \to \pm\infty$ we ask that the wave be outgoing. For the moment consider $x_3 \geq 0$. The potential φ can be represented as

$$\varphi = \frac{1}{(2\pi)^2}\int_{-\infty}^{\infty}\int_{-\infty}^{\infty} {}^*\varphi(k_1, k_2)e^{i(k_1x_1+k_2x_2+k_3x_3)}dk_1\,dk_2, \tag{2.30}$$

with

$$k_3 = \left(k^2 - k_1^2 - k_2^2\right)^{1/2}, \quad \Re(k_3) \geq 0, \quad \Im(k_3) \geq 0. \tag{2.31}$$

This is a solution to (2.27). The term k_3 has been defined so that the wavefield is outgoing. Note that we have moved directly to the step represented previously by (2.24). Applying the condition at $x_3 = 0$, (2.29), gives

$${}^*\varphi(k_1, k_2) = \frac{A}{k}\int_{-\infty}^{\infty}\int_{-\infty}^{\infty}\frac{e^{ik\rho}}{\rho}e^{-i(k_1x_1+k_2x_2)}dx_1\,dx_2. \tag{2.32}$$

Integrals of this form are most readily evaluated by transforming the coordinates of the integrand to polar ones, both in the physical and in the transform space. The transformations are

$$k_1 = \kappa \cos\psi, \quad k_2 = \kappa \sin\psi, \quad x_1 = \rho\cos\theta, \quad x_2 = \rho\sin\theta, \tag{2.33}$$

where $\kappa = (k_1^2 + k_2^2)^{1/2}$ and ρ was given previously. The integral in (2.32) can now be written as

$$^*\varphi = \frac{A}{k}\int_0^{2\pi} d\theta \int_0^{\infty} e^{i\rho[k-\kappa\cos(\psi-\theta)]}\, d\rho. \tag{2.34}$$

Performing the integration gives

$$^*\varphi = 2\pi i A/kk_3, \tag{2.35}$$

where k_3 is given by (2.31). Therefore, the spherical wave, (2.28), can be represented as an integral, taken over both homogeneous and inhomogeneous waves, given by

$$\varphi = \frac{iA}{2\pi k}\int_{-\infty}^{\infty}\int_{-\infty}^{\infty} e^{i(k_1x_1+k_2x_2+k_3|x_3|)}\frac{dk_1\,dk_2}{k_3}. \tag{2.36}$$

This then is the plane-wave representation of a spherical wave. It serves to indicate again the centrality of plane waves as kinematical objects. Note that x_3 has been replaced with its absolute value so as to give a spherical wave over all space. Had we assumed x_3 negative, then the definition of the branch of the function k_3, (2.31), would have to be changed. The outcome of assuming both x_3 negative and changing the branch of k_3 that is taken is equivalent to retaining the definition of (2.31) and replacing x_3 with its absolute value.[3]

It is possible to do still more with this representation. Each plane wave in the integrand has the form $e^{i\boldsymbol{k}\cdot\boldsymbol{x}}$, where $\boldsymbol{k}$ is the complex wavevector and $\boldsymbol{x}$ the position vector. Written this way, it is possible to imagine that a spherical wave could be constructed from a bursting of wavevectors from some compact source region, with each wavevector piercing a spherical surface. It is therefore enough to identify each wavevector with two angles.

Again let us momentarily assume that $x_3 > 0$. Consider the following transformation, sometimes called the *Sommerfeld transformation*.

$$\kappa = k\sin\xi, \qquad k_3 = k\cos\xi, \tag{2.37}$$

[3] There are two points in this calculation that have not been explained clearly. The first, how the branch of a radical such as k_3 is determined, is explained in Section 3.4.4. The second, why we can assume that $e^{ik\rho} \to 0$ as $\rho \to \infty$, is explained in Section 4.4.

where κ was given previously. The variables k_1 and k_2 both vary from $-\infty$ to $+\infty$, so that κ varies from 0 to ∞. Moreover, k_3 must satisfy (2.31). The integration with respect to ψ goes from 0 to 2π. But the integration over ξ is more complicated because ξ must vary over both real and complex values to capture the full range of κ. Let $\xi = \xi_r + i\xi_i$, with ξ_r and ξ_i both real. Then $\sin(\xi_r+i\xi_i) = \sin\xi_r \cosh\xi_i + i\cos\xi_r \sinh\xi_i$ and $\cos(\xi_r+i\xi_i) = \cos\xi_r \cosh\xi_i - i\sin\xi_r \sinh\xi_i$. Let ξ_r vary from 0 to $\pi/2$ and keep $\xi_i = 0$, so that κ varies from 0 to k and $\Re(k_3) \geq 0$. To have κ continue to ∞, we hold $\xi_r = \pi/2$ and let ξ_i vary to either $+\infty$ or $-\infty$. To keep $\Im(k_3) \geq 0$ we must have ξ_i vary from 0 to $-\infty$. We are integrating in the complex ξ plane so that a strict adherence to this contour is not essential, but the contour must begin at zero and end somewhere near $\pi/2 - i\infty$. Further, it must be such that the integral is convergent and represents an outgoing wavefield. Carefully changing the variables of integration in (2.36) gives

$$\varphi = \frac{iA}{2\pi k} \int_0^{2\pi} d\psi \int_0^{\pi/2-i\infty} e^{i\mathbf{k}\cdot\mathbf{x}} \sin\xi \, d\xi. \tag{2.38}$$

Note that k_3 has has been removed by this transformation. The expression (2.38) is called an angular-spectrum representation and succinctly captures the image of a spherical wave as a burst of wavevectors. Note that to fully achieve the spherical wavefront, inhomogeneous waves are also needed.

2.3.3 An Angular-Spectrum Representation of a Cylindrical Wave

To achieve a plane-wave or angular-spectrum representation of a cylindrical wave, it is easier to work from the equation of motion with a line force than to try to pose a signaling problem. This is done in *Problem 4.1*. However, as indicated in the Appendix, an outgoing cylindrical wave is described by the Hankel function, $H_0^{(1)}(k\rho)$. Sommerfeld (1964) describes how to represent this function as an angular spectrum and gives the result

$$H_0^{(1)}(k\rho) = \frac{1}{\pi} \int_C e^{ik\rho\cos\xi} \, d\xi. \tag{2.39}$$

The contour C begins at $-\eta + i\infty$ and ends at $\eta - i\infty$, where $\eta \in (0, \pi)$.

2.4 Asymptotic Ray Expansion

A much older way to represent waves is by using rays. Born and Wolf (1986) give a concise introduction to ray theory for electromagnetic waves, while Achenbach et al. (1982) provide a detailed asymptotic development of elastic-wave ray theory. There are, at least, two ways to approach ray descriptions. The one presented next builds upon the time-harmonic plane wave, with the assumption, lurking in the background, that we can synthesize the time-dependent case from this construction, should we need to. However, it is also possible to use the fact that hyperbolic equations permit discontinuities to be transported along their characteristics. A wave is permitted to have a discontinuity at its wavefront and the kinematics of the discontinuity is developed. This is the approach taken by Hudson (1980) for elastic waves, and in a more general context by Whitham (1974). The two approaches are connected by the fact that a discontinuity in a wavefront engages predominantly the high-frequency components of the temporal Fourier transform (Lighthill, 1978). We expect ray descriptions to be useful only when the wavelengths are small with respect to any other characteristic length in the propagation environment.

2.4.1 Compressional Wave

We begin by working with the scalar potential φ. Knowing the representation for φ, it is straightforward to construct that for the vector potential ψ and hence that for the particle displacement $\boldsymbol{u}$. We are, then, left examining the equation

$$\nabla^2\varphi + k^2\varphi = 0, \tag{2.40}$$

where we have again written c for c_L, so that $k = \omega/c$.

As a generalization of a plane wave, we assume that φ can be expanded as

$$\varphi(\boldsymbol{x}) \sim e^{ik[S(\boldsymbol{x})-ct]}(ik)^{-\alpha-1}\sum_{n=0}^{\infty}(ik)^{-n}A_n(\boldsymbol{x}), \tag{2.41}$$

where $\alpha < 1$ is a positive number (the term raised to the power α is added for generality). We have added the $e^{-i\omega t}$, where t is time, to the exponential to aid in our subsequent discussions. In writing this expression, we are using two ideas. First, we note that any surface can be approximated by its tangent planes and hence locally any wavefield will have a propagating part or phase[4] that behaves as does that of a plane wave. Second, by examining the representations (2.38) and (2.39), we should expect that the amplitude of a wave with a curved wavefront depends in some way on k or, equivalently, the wavelength $\lambda = 2\pi/k$.

[4] The word *phase* has a rather confused meaning in wave propagation. Looking back at (2.10), we note the argument of the amplitude A makes a contribution to the exponential term of a plane wave. This contribution is sometimes called the phase. However, referring to (2.41), we note that the exponential, $kS(\boldsymbol{x})$, is also called the phase. When the word *phase* is used in this book, it refers to this latter term, that controlled by the wavenumber k.

Therefore, it is reasonable that the amplitude can be expanded as an asymptotic power series[5] in the principal parameter λ, or equivalently k^{-1}, and that the approximation becomes increasingly accurate as $\lambda \to 0$ or $k \to \infty$. We begin the expansion with $k^{-\alpha-1}$ so that the approximation for the particle displacement will start as $k^{-\alpha}$. Lastly, it is not good workmanship to expand in a parameter with a dimension because the measure of large or small remains too vague. In the present case the reader should imagine that k is multiplied by a distance representing that between interactions, call it L, and that it is kL that is very large. Rather than introduce an extra parameter, however, we suppress the L unless explicitly needed.

Substituting (2.41) into (2.40) gives

$$\sum_{n=0}^{\infty} \Big\{(\nabla S \cdot \nabla S - 1)A_n + [2(\nabla S \cdot \nabla)A_{n-1} + \nabla^2 S A_{n-1}] + \nabla^2 A_{n-2}\Big\}(ik)^{-(n-1)} = 0, \tag{2.42}$$

where the A_n with negative subscripts are zero. The set $\{(ik)^{-n}\}$ is linearly independent. Accordingly, for $A_0 \neq 0$,

$$\nabla S \cdot \nabla S = 1. \tag{2.43}$$

Note that $|\nabla S| = 1$. This equation is called the eikonal equation. At $n = 1$ we have

$$2(\nabla S \cdot \nabla)A_0 + (\nabla^2 S)A_0 = 0, \tag{2.44}$$

and, for $n > 1$,

$$2(\nabla S \cdot \nabla)A_{n-1} + (\nabla^2 S)A_{n-1} = -\nabla^2 A_{n-2}. \tag{2.45}$$

These last two equations are called the transport equations. Collectively, (2.43)–(2.45) provide a recursive scheme whereby the solution to one equation gives the missing information needed for the solution to the next. The solution of the eikonal equation gives a family of curves along which the amplitudes A_n are transported.

We define a ray as the vector $k\nabla S$. It is tangent to one of the curves found from solving (2.43). We define a wavefront, just as we did following (2.1),

[5] Holmes (1995) and Hinch (1991) provide useful starting points for the reader to learn more about asymptotic and perturbation methods.

as the family of surfaces $S(\boldsymbol{x}) = ct + C$, where t is a parameter giving each member of the family and C is a constant.

To solve (2.43), let $\boldsymbol{x}(s, \boldsymbol{q})$ define a family of curves that are orthogonal to each surface $S(\boldsymbol{x}) = ct + C$. The variable s defines the arclength along the curve, while $\boldsymbol{q}$ indicates each member of the family. For the moment, we imagine that $\boldsymbol{x}$ does not depend on $\boldsymbol{q}$ and suppress any reference to it until we need it. By definition,

$$d\boldsymbol{x}/ds = \nabla S. \tag{2.46}$$

Setting $\hat{\boldsymbol{p}} = \nabla S$, we see that the left-hand side of (2.43) is $\hat{\boldsymbol{p}} \cdot \nabla S$, which is nothing more than the directional derivative dS/ds. Thus $S(\boldsymbol{x}) = s(\boldsymbol{x}) + ct_0$, where t_0 is an initial time. We have assumed that the equation for each curve $\boldsymbol{x}(s)$ can be inverted to give s as a function of $\boldsymbol{x}$. We do not at present know $\boldsymbol{x}(s)$.

Direct differentiation and using (2.43) indicate that $d\hat{\boldsymbol{p}}/ds = 0$ along a curve $\boldsymbol{x}(s, \boldsymbol{q})$, where we have reinserted the $\boldsymbol{q}$ to indicate that we are now considering a family (a pencil) of such curves. Therefore the curves are straight lines whose equations are

$$\boldsymbol{x}(s, \boldsymbol{q}) = \boldsymbol{x}_0(\boldsymbol{q}) + s\hat{\boldsymbol{p}}(\boldsymbol{q}). \tag{2.47}$$

In Fig. 2.2 we have sketched some aspects of the construction. The vector $\boldsymbol{x}_0(\boldsymbol{q})$ is the starting point for a ray identified by $\boldsymbol{q}$ on the initial wavefront $S(\boldsymbol{x}_0) = ct_0$. Both these equations are assumed known or given. The components of the vector $\boldsymbol{q} = (q_1, q_2)$ constitute a coordinate system on the initial wavefront and define the point at which the ray is launched at $t = t_0$. The rays propagate along and are tangent to $\boldsymbol{x}(s, \boldsymbol{q})$ and pierce each wavefront perpendicularly. Inverting (2.47) to obtain $s(\boldsymbol{x})$ and $\boldsymbol{q}(\boldsymbol{x})$ gives the equation for a wavefront as $S(\boldsymbol{x}) = s(\boldsymbol{x}) + S(\boldsymbol{x}_0)$, where $s(\boldsymbol{x}) = c(t - t_0)$[6]. Recall that (2.47), by the implicit function theorem, is locally invertible if the Jacobian does not vanish. Surfaces at which this Jacobian vanishes are called caustics. In the neighborhood of these surfaces, the expansion (2.41) becomes disordered. Choi and Harris (1989, 1990) give an interesting example of caustic formation in elastic materials, though their asymptotic analysis starts from integral representations, rather than ray expansions.

[6] Note that this description of the rays, while qualitatively correct for anisotropic or inhomogeneous materials, requires some substantial modification in these cases.

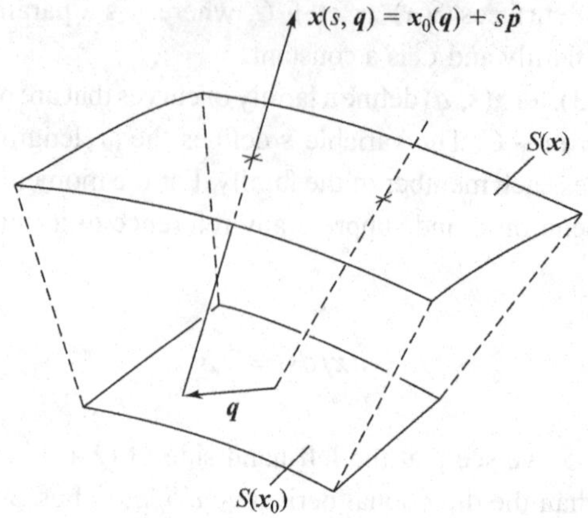

Fig. 2.2. A ray is launched in a direction normal to an initial wavefront $S(\boldsymbol{x}_0)$ from the point $\boldsymbol{q}$ on $S(\boldsymbol{x}_0)$. It is a straight line. After propagating a distance s along this line, it pierces the wavefront $S(\boldsymbol{x})$, again in a normal direction. The starting point of the ray is $\boldsymbol{x}_0(\boldsymbol{q})$, and its present position is $\boldsymbol{x}(s, \boldsymbol{q}) = \boldsymbol{x}_0(\boldsymbol{q}) + s\hat{\boldsymbol{p}}$, where $\hat{\boldsymbol{p}}$ is a unit vector pointing along (tangent to) the ray

Weatherburn (1939)[7] shows that

$$\nabla^2 S = 1/\rho_1 + 1/\rho_2, \tag{2.48}$$

where ρ_1 and ρ_2 are the principal radii of curvature of the surface $S(\boldsymbol{x}) = s(\boldsymbol{x}) + S(\boldsymbol{x}_0)$. These radii take a sign. If a wavefront is expanding, its radii of curvature are positive, and the normal to the surface points in the direction of expansion. If the wavefront is contracting, the radii of curvature are negative, but the normal continues to point in the direction of propagation, namely that of contraction. Moreover, because s is the distance along the straight lines orthogonal to each surface, $\rho_1 = \rho_{01} + s$ and $\rho_2 = \rho_{02} + s$, where the ρ_{0i} are the radii of curvature of the initial surface $S(\boldsymbol{x}_0) = ct_0$. Accordingly, the first transport equation, (2.49), can be written as

$$\frac{dA_0}{ds} + \frac{1}{2}\left(\frac{1}{\rho_{01} + s} + \frac{1}{\rho_{02} + s}\right) A_0 = 0. \tag{2.49}$$

This is a differential equation along each ray $\boldsymbol{q}$. Its solution is

$$A_0 = \frac{A(\boldsymbol{q})}{[(\rho_{01} + s)(\rho_{02} + s)]^{1/2}}, \tag{2.50}$$

where $A(\boldsymbol{q})$ is assumed to be given.

[7] The difference in sign from that in Weatherburn arises because of how we have defined the signs of the radii of curvature.

Collecting the various pieces and noting that the compressional component of the particle displacement $\boldsymbol{u}_L = \nabla\varphi$, gives, to leading order,

$$\boldsymbol{u}_L \sim \frac{\hat{\boldsymbol{p}} A_L(\boldsymbol{q})}{(ik_L)^{\alpha}[(\rho_{01}+s)(\rho_{02}+s)]^{1/2}} e^{ik_L[s(\boldsymbol{x})-c_L(t-t_0)]}, \tag{2.51}$$

where we have put back the subscript L to indicate that we have calculated the compressional component. The power α must also be given, typically as part of the initial data for the problem.

This is a remarkable expression. It captures our intuitive notions of a wave as an object whose phase steadily increases as we move along a curve, the tangent vector to the curve being a ray, and whose amplitude decays geometrically such as to conserve energy (*Problem 2.4*). Note also that the denominator may vanish, in which case our asymptotic approximation is no longer accurate. This happens on the caustic surfaces mentioned previously. Lastly, note that the polarization of the leading-order term (2.51) is that of a longitudinal wave, but this does not preclude the possibility that the higher-order terms could have different polarizations.

Problem 2.4 Conservation of Energy in a Ray Tube

Show, by multiplying through by A_0, that the first transport equation, (2.49), can be written as

$$\nabla \cdot \left(\nabla S A_0^2\right) = 0. \tag{2.52}$$

Integrate this over a tube formed from rays and, for an infinitesimal cross section, deduce the result, (2.50). Interestingly, this equation establishes more than just the conservation of time-averaged energy. The A_0 can be complex and the solution to (2.52) gives both its magnitude and its argument. If conservation of energy were our only goal, we should consider $\nabla \cdot (\nabla S A_0 A_0^*) = 0$. The reader is encouraged to do so.

2.4.2 Shear Wave

Next we consider a shear wave. The vector potential $\boldsymbol{\psi}$ is written as

$$\boldsymbol{\psi}(\boldsymbol{x}) \sim e^{ik_T[S(\boldsymbol{x})-c_T t]}(ik_T)^{-\alpha-1}\sum_{n=0}^{\infty}(ik_T)^{-n}\boldsymbol{A}_n(\boldsymbol{x}), \tag{2.53}$$

where $k_T = \omega/c_T$. Recall that $\nabla \cdot \boldsymbol{\psi} = 0$. Provided we do not seek terms beyond the leading one, this implies that

$$\boldsymbol{A}_0 \cdot \nabla S = 0. \tag{2.54}$$

The kinematic analysis here will be identical to that of the previous section so that $\hat{\boldsymbol{p}} = \nabla S$ gives the direction of the rays, which propagate along straight lines. Hence, $\boldsymbol{A}_0$ will be orthogonal to $\hat{\boldsymbol{p}}$ and have two linearly independent components, $A_0^{TV}\hat{\boldsymbol{d}}_{TH}$ and $A_0^{TH}\hat{\boldsymbol{d}}_{TV}$. We select the unit vectors $\hat{\boldsymbol{d}}_I$ so that

$$\hat{\boldsymbol{p}} \wedge \hat{\boldsymbol{d}}_{TV} = \hat{\boldsymbol{d}}_{TH}. \tag{2.55}$$

At this point the analysis given in the previous section applies to each component of $\boldsymbol{A}_0$. We find that, to leading order, the shear particle displacement is

$$\boldsymbol{u}_I \sim \frac{\chi_I \hat{\boldsymbol{d}}_I A_I(\boldsymbol{q})}{(ik_T)^\alpha[(\rho_{01}+s)(\rho_{02}+s)]^{1/2}} e^{ik_T[s(\boldsymbol{x})-c_T(t-t_0)]}, \tag{2.56}$$

where $\chi_I = \mp 1$ as $I = TH, TV$. Note that the radii of curvature ρ_{0i} are different for the two wave types (2.51) and (2.56). It takes the presence of a source or boundary to couple the waves together. As with the compressional wave, the polarization can change as one proceeds to the higher-order terms.

Appendix: Spherical and Cylindrical Waves

The following is a summary of information about spherical and cylindrical waves.

To begin, we consider a wavefield whose only dependence is upon the radial coordinate r and whose only component of particle displacement u is in the radial direction. The equation of motion becomes

$$\frac{\partial^2 u}{\partial r^2} + \frac{2}{r}\frac{\partial u}{\partial r} - \frac{2u}{r^2} = \frac{1}{c_L^2}\frac{\partial^2 u}{\partial t^2}. \tag{2.57}$$

Such a wavefield could be excited by a uniformly pressurized spherical cavity. Setting $u = \partial\varphi/\partial r$, we find that this equation reduces to

$$\frac{\partial^2(r\varphi)}{\partial r^2} = \frac{1}{c_L^2}\frac{\partial^2(r\varphi)}{\partial t^2}, \tag{2.58}$$

and its solution is

$$\varphi(r,t) = \frac{1}{r} f\left(t - \frac{r}{c_L}\right) + \frac{1}{r} g\left(t + \frac{r}{c_L}\right), \tag{2.59}$$

where f and g are arbitrary functions. The first term represents an outgoing wave and the second an incoming one. Let us assume that there is only an outgoing wave and take its temporal transform. Then we get

$$\bar{\varphi} = \frac{\bar{f}(\omega)}{r} e^{ik_L r}, \tag{2.60}$$

in agreement with (2.28). Note that the particle displacement $u = \partial\varphi/\partial r$ so that there are both r^{-1} and r^{-2} terms.

Next we consider inplane, rotary shear motion. We use a cylindrical coordinate system to describe it. The only component of particle displacement is v, a displacement in the angular, θ direction. The only dependence is upon the polar, radial coordinate ρ. There is no dependence on x_3. The equation of motion is

$$\frac{\partial^2 v}{\partial \rho^2} + \frac{1}{\rho}\frac{\partial v}{\partial \rho} - \frac{v}{\rho^2} = \frac{1}{c_T^2}\frac{\partial^2 v}{\partial t^2}. \tag{2.61}$$

Such a wavefield could be excited by a traction, solely in the θ direction, on the walls of a cylindrical cavity. Setting $v = -\partial\psi_3/\partial\rho$, we find that this equation reduces to

$$\frac{\partial^2 \psi_3}{\partial \rho^2} + \frac{1}{\rho}\frac{\partial \psi_3}{\partial \rho} = \frac{1}{c_T^2}\frac{\partial^2 \psi_3}{\partial t^2}. \tag{2.62}$$

Taking the temporal transform, we get

$$\frac{\partial^2 \bar{\psi}_3}{\partial \rho^2} + \frac{1}{\rho}\frac{\partial \bar{\psi}_3}{\partial \rho} + k_T^2 \bar{\psi}_3 = 0. \tag{2.63}$$

This is Bessel's equation of order zero and has, as two of its linearly independent solutions, the Hankel functions $H_0^{(1,2)}(k_T\rho)$. The asymptotic behavior of these functions is

$$H_0^{(1,2)}(k_T\rho) \sim (2/\pi k_T\rho)^{1/2} e^{\pm i(k_T\rho - \pi/4)}, \qquad k_T\rho \to \infty, \tag{2.64}$$

where the plus sign goes with the 1 and the minus sign with the 2. From this behavior, we see that $H_0^{(1)}(k_T\rho)$ represents an outgoing wave and $H_0^{(2)}(k_T\rho)$ an incoming one. Recall that the particle displacement is $v = -\partial\psi_3/\partial\rho$.

References

Achenbach, J.D., Gautesen, A.K., and McMaken, H. 1982. *Ray Methods for Waves in Elastic Solids*, pp. 77–88. Boston: Pitman.

Born, M. and Wolf, E. 1986. *Principles of Optics*, 6th (corrected) ed., pp. 109–127. Oxford: Pergamon.

Boulanger, Ph. and Hayes, M. 1993. *Bivectors and Waves in Mechanics and Optics.* New York: Chapman and Hall.

Choi, H.C. and Harris, J.G. 1989. Scattering of an ultrasonic beam from a curved interface. *Wave Motion* **11**: 383–406.

Choi, H.C. and Harris, J.G. 1990. Focusing of an ultrasonic beam by a curved interface. *Wave Motion* **12**: 497–511.

Clemmow, P.C. 1966. *The Plane Wave Spectrum Representation of Electromagnetic Fields.* Oxford: Pergamon.

Hinch, E.J. 1991. *Perturbation Methods.* Cambridge: University Press.

Holmes, M.H. 1995. *Introduction to Perturbation Methods.* New York: Springer.

Hudson, J.A. 1980. *The Excitation and Propagation of Elastic Waves.* New York: Cambridge.

Lighthill, M.J. 1978. *Fourier Analysis and Generalized Functions,* pp. 46–57. New York: Cambridge

Sommerfeld, A. 1964. *Partial Differential Equations in Physics,* Lectures on Theoretical Physics, Vol. VI, pp. 84–101. Translated by E.G. Straus. New York: Academic.

Weatherburn, C.E. 1939. *Differential Geometry of Three Dimensions,* Vol. 1, pp. 225–227. Cambridge: University Press.

Whitham, G.B. 1974. *Linear and Nonlinear Waves.* New York: Wiley-Interscience.

3

Reflection, Refraction, and Interfacial Waves

Synopsis

Chapter 3 describes reflection and refraction at an interface between two materials having different densities and wavespeeds. Moreover, the chapter describes waves that propagate along an interface, while decaying perpendicularly away from it. These waves, in many cases, must be continuously renewed from a wavefield in the interior of one or both the materials to sustain their propagation. However, a traction-free elastic surface is special in that a wave can be excited on such a surface and not require an interior wavefield to sustain it.

There is an intimate relation between the topography of the complex wavenumber plane and the physical manifestation of the propagating wavefield, as we have previously noted. Extending this connection, we discuss how branches of the function $(k^2 - z^2)^{1/2}$, where z is the complex variable and k a parameter, are selected and how these selections manifest themselves in the physical domain.

3.1 Reflection of a Compressional Plane Wave

We consider a longitudinal or compressional plane wave incident to a traction-free surface. Figure 3.1 indicates the geometry of the problem along with a schematic representation of the interaction. Further, we assume that the disturbance is time harmonic, but suppress the $e^{-i\omega t}$ unless it is explicitly needed. We do so knowing that using (1.36) will carry the resulting expressions into corresponding time-dependent ones. The incident wave $\boldsymbol{u}_0$ is described by

$$\boldsymbol{u}_0 = A_0\,\hat{\boldsymbol{p}}_0 e^{ik_L \hat{\boldsymbol{p}}_0\cdot\boldsymbol{x}}, \tag{3.1}$$

where the unit vector $\hat{\boldsymbol{p}}_0$, which describes both the polarization of the wave and its direction of propagation, is given by

$$\hat{\boldsymbol{p}}_0 = \sin\theta_0\,\hat{\boldsymbol{e}}_1 + \cos\theta_0\,\hat{\boldsymbol{e}}_2. \tag{3.2}$$

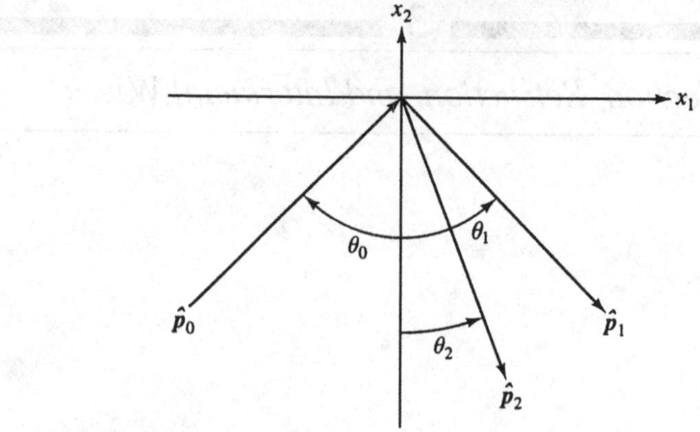

Fig. 3.1. The elastic solid fills the $x_2 < 0$ half-space, while $x_2 > 0$ is a vacuum. The surface is free of tractions. A compressional plane wave is incident to the surface in the direction $\hat{\boldsymbol{p}}_0$.

The angle θ_0 is indicated in Fig. 3.1, and the $\hat{\boldsymbol{e}}_i$ are the unit vectors along the x_i axes ($i = 1, 2$). The amplitude A_0 is real and positive. As in previous work, $k_L = \omega / c_L$. The incident polarization has no antiplane component. Therefore any waves excited at the boundary must also be inplane. Further, the incident wave is plane so that it imposes a projection of its phase everywhere on the surface. We therefore expect both compressional and shear plane waves to be reflected[1] from the boundary, because both wave types can have matching phase components along the surface.

The reflected compressional wave is described by

$$\boldsymbol{u}_1 = A_1 \hat{\boldsymbol{p}}_1 e^{ik_L \hat{\boldsymbol{p}}_1 \cdot \boldsymbol{x}}, \tag{3.3}$$

where the vector $\hat{\boldsymbol{p}}_1$ is given by

$$\hat{\boldsymbol{p}}_1 = \sin\theta_1 \hat{\boldsymbol{e}}_1 - \cos\theta_1 \hat{\boldsymbol{e}}_2. \tag{3.4}$$

The reflected shear wave is described by

$$\boldsymbol{u}_2 = A_2 \hat{\boldsymbol{d}}_2 e^{ik_T \hat{\boldsymbol{p}}_2 \cdot \boldsymbol{x}}, \tag{3.5}$$

where the propagation direction is given by

$$\hat{\boldsymbol{p}}_2 = \sin\theta_2 \hat{\boldsymbol{e}}_1 - \cos\theta_2 \hat{\boldsymbol{e}}_2, \tag{3.6}$$

[1] I use the phrase *scattered wave* or *scattered wavefield* to describe almost any disturbance that is returned from or perturbed by an obstacle struck by an incident wave. When the obstacle is impenetrable, the scattered wavefields generally break into reflected and diffracted ones. Any wavefield that penetrates a shadowed region is a diffracted one; otherwise, it is reflected.

but the polarization direction by

$$\hat{\boldsymbol{d}}_2 = \hat{\boldsymbol{e}}_3 \wedge \hat{\boldsymbol{p}}_2. \tag{3.7}$$

Figure 3.1 defines the angles θ_1 and θ_2. The amplitudes A_1 and A_2 may be complex. And as in previous work, $k_T = \omega/c_T$. Note that both reflected waves propagate away from the surface. Also note that the polarization of the reflected shear wave is chosen so that $\hat{\boldsymbol{p}}_2 \wedge \hat{\boldsymbol{d}}_2 = \hat{\boldsymbol{e}}_3$.

3.1.1 Phase Matching

The boundary conditions at $x_2 = 0$ are that $\tau_{22} = \tau_{21} = 0$. Direct application of these boundary conditions can be tedious. However, a little thought indicates that they must take the form

$$(\cdots)A_0 e^{ik_L \sin\theta_0 x_1} + (\cdots)A_1 e^{ik_L \sin\theta_1 x_1} + (\cdots)A_2 e^{ik_T \sin\theta_2 x_1} = 0. \tag{3.8}$$

This condition must hold for all x_1. This can happen only if $\theta_0 = \theta_1$ and $s_L \sin\theta_0 = s_T \sin\theta_2$, where $s_I = c_I^{-1}$ is the slowness. These two conditions are called the *phase-matching* condition. The phrase *phase-matching* is often used as a verb in the sense that "the reflected waves phase match to the incident one." Phase matching is nothing more than a demand that the projections, on the surface $x_2 = 0$, of the various wavelengths be equal or, equivalently, because ω is common to all the k_I, that the wavespeeds along the surface be identical. In fact it is a kinematical condition, rather than a kinetic one. The surface is translationally invariant and the disturbance at one point differs from that at another only by an exponential phase term that indicates propagation, in this case to the right. This is a similar condition to that used to derive (1.66). Phase matching is a fundamental principle of linear wave propagation, and, whenever it occurs, something physically interesting will happen.

There is a very useful diagram that geometrically represents the phase-matching condition. We define the slowness vectors

$$\boldsymbol{s}_0 = \hat{\boldsymbol{p}}_0/c_L, \qquad \boldsymbol{s}_1 = \hat{\boldsymbol{p}}_1/c_L, \qquad \boldsymbol{s}_2 = \hat{\boldsymbol{p}}_2/c_T. \tag{3.9}$$

The tip of each slowness vector describes a circle, a slowness surface,[2] as the parameter θ_0 is varied through 2π. This is shown in Fig. 3.2. The phase-matching condition is stated as the condition that the s_1 components of each slowness vector must be equal. We have indicated this by the vertical dashed line. In the

[2] Slowness surfaces for isotropic materials are always spheres (circles really, because only two dimensions are involved), but for anisotropic materials they can become very elaborate indeed. Auld (1990) gives an account of these surfaces and of their use in determining phase-matching.

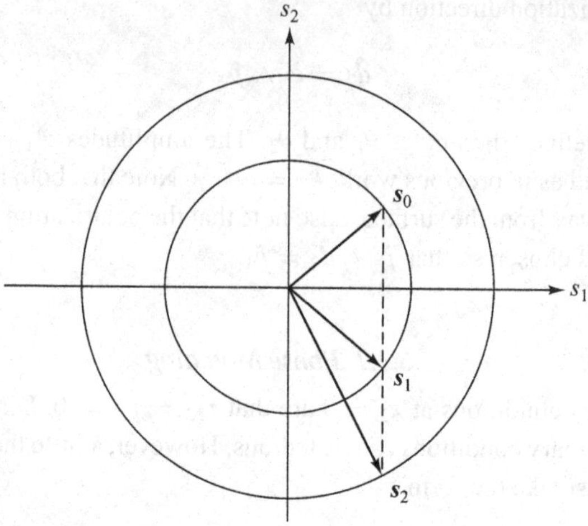

Fig. 3.2. The slowness surfaces for an isotropic elastic solid. The phase-matching condition is indicated geometrically by demanding that the horizontal components (projections on the surface $x_2 = 0$) of the slowness vectors be equal.

case we are examining here, for real θ_0, θ_2 must be real and less than $\pi/2$, even in the limit that $\theta_0 = \pi/2$.

3.1.2 Reflection Coefficients

Problem 3.1 asks the reader to calculate the longitudinal or compressional reflection coefficient $R_L(\theta_0) = A_1/A_0$ and the transverse or shear reflection coefficient $R_T(\theta_0) = A_2/A_0$. When this is done the reader will find that

$$R_L(\theta_0) = A_-(\theta_0)/A_+(\theta_0), \tag{3.10}$$

$$R_T(\theta_0) = 2\kappa \sin 2\theta_0 \cos 2\theta_2 / A_+(\theta_0). \tag{3.11}$$

The auxiliary functions are

$$A_\mp = \sin 2\theta_0 \sin 2\theta_2 \mp \kappa^2 \cos^2 2\theta_2. \tag{3.12}$$

These terms must be supplemented with the condition $s_L \sin\theta_0 = s_T \sin\theta_2$. The term $\kappa = c_L/c_T$ is given by

$$\kappa = [2(1-\nu)/(1-2\nu)]^{1/2}, \tag{3.13}$$

where ν is Poisson's ratio. For many materials, κ is close to $3^{1/2}$.

Note that, provided θ_0 is real, these reflection coefficients have no frequency dependence and hence can be used both for time-harmonic and time-dependent plane waves.

Problem 3.1 Reflection Coefficients

Show that applying the boundary condition $\tau_{22} = 0$ leads to

$$(\lambda + 2\mu \cos^2 \theta_0)(A_1/A_0) - \kappa\mu \sin 2\theta_2(A_2/A_0) = -(\lambda + 2\mu \cos^2 \theta_0), \tag{3.14}$$

and that applying the condition $\tau_{21} = 0$ leads to

$$-\mu \sin 2\theta_0(A_1/A_0) - \kappa\mu \cos 2\theta_2(A_2/A_0) = -\mu \sin 2\theta_0. \tag{3.15}$$

Hence deduce (3.10)–(3.12).

Problem 3.2 Conservation of Energy

Show that the power flux into the surface equals that away from it. That is, show that

$$|R_L|^2 + |R_T|^2 \frac{\cos\theta_2}{\kappa \cos\theta_0} = 1. \tag{3.16}$$

Hint. The calculation becomes straightforward once the reader realizes that he or she must be careful to use the same element of area at the surface to calculate both the flux into the surface and that away from it.

While there are several other problems of this general kind that could be considered, we treat only one other, namely the reflection and refraction of an antiplane shear wave at the interface between two contrasting materials.

3.2 Reflection and Refraction

We consider a plane, antiplane shear wave incident to an interface between two materials that are in welded contact. In this case both the traction and the particle displacement are continuous across the interface. To simplify the notation, we replace c_T, k_T and similarly labeled symbols by c, k, and so on. It is clear that only an antiplane particle displacement can be excited at the interface by an

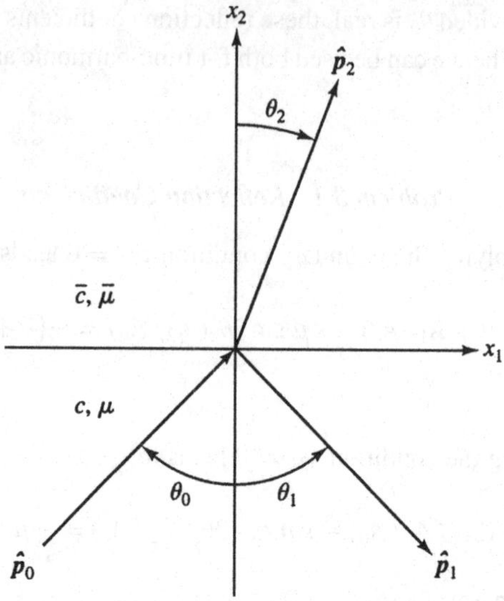

Fig. 3.3. In the $x_2 < 0$ half-space the density is ρ, the elastic shear modulus μ, and the wavespeed $c = (\mu/\rho)^{1/2}$, while in the $x_2 > 0$ half-space the density is $\bar{\rho}$, the elastic shear modulus $\bar{\mu}$, and the wavespeed $\bar{c} = (\bar{\mu}/\bar{\rho})^{1/2}$. A plane, antiplane shear wave is incident to the surface in the direction $\hat{p}_0$.

incident wave with an antiplane polarization, so that the scattered waves must also be similarly polarized.

Figure 3.3 indicates the various angles. The incident wave is described by

$$u_{30} = A_0 e^{ik\hat{p}\cdot x}, \tag{3.17}$$

where the subscript 3 indicates the component and therefore the polarization, and the 0 that the wave is incident. The incident direction is given by (3.2). The reflected wave, indicated with the additional subscript 1, is described by

$$u_{31} = A_1 e^{ik\hat{p}\cdot x}, \tag{3.18}$$

with its propagation direction given by (3.4). The wave that is transmitted across the interface is said to be refracted. The refracted wave, indicated by the additional subscript 2, is described by

$$u_{32} = A_2 e^{i\bar{k}\hat{p}_2\cdot x}, \tag{3.19}$$

with its propagation direction given by

$$\hat{p}_2 = \sin\theta_2 \hat{e}_1 + \cos\theta_2 \hat{e}_2. \tag{3.20}$$

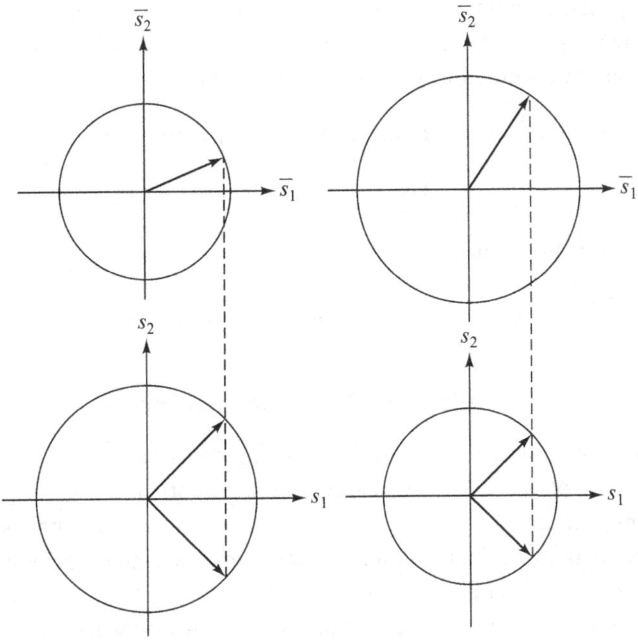

Fig. 3.4. The upper slowness surface is that for the material of the upper half-space in Fig. 3.3 and the lower for the material of the lower half-space. The left-hand side indicates the case $\bar{c} > c$, and the right the opposite one. The phase-matching condition is indicated by the dashed vertical lines.

In Fig. 3.4 the slowness surfaces for the two materials are arranged vertically in a pattern corresponding with the positions of the half-spaces in Fig. 3.3. The two possible cases are shown. The left-hand side indicates that when the upper material is faster and the right-hand side that when it is slower. Phase matching demands that the s_1 components of the slowness vectors be equal; that is, $\theta_0 = \theta_1$ and $\bar{s}\sin\theta_2 = s\sin\theta_0$. More importantly, the slowness diagrams indicate the occurrence of critical refraction. This occurs when a wave incident to the interface from the slower material excites a wave skimming along the interface in the faster material. In principle the reverse can also occur. For the case $\bar{c} > c$ or, equivalently $\bar{s} < s$, the critical angle of incidence is θ_{0c}, where $s\sin\theta_{0c} = \bar{s}$. Of course θ_0 can be greater than θ_{0c}, in which case θ_2 must be complex to satisfy the phase-matching condition. This gives rise to an inhomogeneous plane wave in the upper material but to no time-average flux of energy across the interface, as the solution to *Problem 3.4* indicates. This wave, clinging to the interface, is propelled along it by the waves in the lower material.

Applying the continuity of traction and particle displacement at the interface gives the antiplane shear, reflection coefficient $R(\theta_0)$ and the antiplane shear, refraction or transmission coefficient $T(\theta_0)$, namely

$$R(\theta_0) = C_-(\theta_0)/C_+(\theta_0), \tag{3.21}$$

$$T(\theta_0) = 2\cos\theta_0/C_+(\theta_0). \tag{3.22}$$

The auxiliary functions are

$$C_\mp = \cos\theta_0 \mp (\bar{\mu}c/\mu\bar{c})\cos\theta_2. \tag{3.23}$$

Problem 3.3 Slowness Surfaces

Figure 3.4 shows how we can place slowness surfaces above one another to work out the phase-matching conditions and the various critical angles. Draw diagrams, similar to Fig. 3.4, of the slowness surfaces for inplane motion. Consider both an incident compressional plane wave and inplane shear one. Consider all the possibilities for critical reflection and critical refraction. Critical reflection, analogously to critical refraction, arises when a wave skimming along the interface or surface of a material is excited by a slower wave incident to the interface from the same material. The reader may be surprised by the large number of occurrences of critical reflection and refraction.

3.3 Critical Refraction and Interfacial Waves

Our discussion of critical refraction indicates that we do not need to consider the angle of refraction as real. Moreover, as *Problem 3.3* has indicated, critical phenomena occur frequently in elastic-wave scattering. We must then regard reflection and transmission coefficients as functions of a complex variable. In fact, we shall find in Chapters 5 and 6 that where the complex plane is punctured by the poles or cut by the branches of these coefficients, interesting wave processes occur.

Let us continue to consider the problem of Section 3.2 and work with the case of $\bar{c} > c$. When $\theta_0 > \theta_{0c}$, θ_2 must be complex to permit $\sin\theta_2 > 1$. To find its form we set $\theta_2 = \pi/2 \pm i\beta$, where β is real and positive. In this case $\sin\theta_2 = \cosh\beta$ and $\cos\theta_2 = \mp i\sinh\beta$. Next we look back at (3.21)–(3.23) and note that $|R(\theta_0)| = 1$, implying that $R(\theta_0) = e^{-i2\varphi}$ and $T(\theta_0) = |T(\theta_0)|e^{-i\varphi}$. Here φ is the argument of $C_+(\theta_0)$. Setting $A_0 = 1$, we find that the particle displacement in $x_2 > 0$ becomes

$$u_{32} = |T(\theta_0)|e^{-i\varphi}e^{\pm\bar{k}\sinh\beta x_2}e^{i\bar{k}\cosh\beta x_1}, \tag{3.24}$$

where $\hat{\boldsymbol{p}}_2$ has been written in full as $\hat{\boldsymbol{p}}_2 = \cosh\beta\,\hat{\boldsymbol{e}}_1 \mp i\sinh\beta\,\hat{\boldsymbol{e}}_2$. Assuming that $\bar{k}$ is positive, we must select $\theta_2 = \pi/2 - i\beta$ to ensure that the wavefield decays as $x_2 \to \infty$. Thus (3.24) is an example of both an inhomogeneous plane wave and of a wave that clings to the interface. We call the wave an interfacial one. However, note that the wave would not exist unless a plane wave were continuously incident to the interface from $x_2 < 0$ sustaining it.

There remains one subtle point. In choosing the minus sign we assumed that $\bar{k}$ and thus ω was positive. If ω were negative then we must choose $\theta_2 = \pi/2 + i\beta$ to ensure decay. Therefore, when $\theta_0 \geq \theta_{0c}$, we must write

$$R(\theta_0) = e^{-i2\varphi\,\mathrm{sgn}\omega}, \qquad T(\theta_0) = |\,T(\theta_0)|e^{-i\varphi\,\mathrm{sgn}\omega}, \tag{3.25}$$

where

$$\mathrm{sgn}\,\omega = \begin{cases} 1, & \omega > 0, \\ -1, & \omega < 0, \end{cases} \tag{3.26}$$

to take account of this ω dependence. The argument φ is determined when ω is positive. *It was to handle such situations that we rewrote the inverse temporal Fourier transform in the form of* (1.36).

Problem 3.4 Flux of Energy in the Upper Material

For the case of critical refraction, show that the time-average flux of energy in $x_2 > 0$ is given by

$$\langle \mathcal{F} \rangle = \tfrac{1}{2}\bar{\mu}\omega\bar{k}|\,T(\theta_0)|^2 e^{-2\bar{k}\,\mathrm{sgn}\omega\,\sinh\beta\,x_2}\cosh\beta\;\hat{\boldsymbol{e}}_1. \tag{3.27}$$

Thus there is no flux of energy to $x_2 \to \infty$ and energy is continually returned to the slower material.

When $\theta_0 < \theta_{0c}$ the reflection and transmission coefficients have no frequency dependence and therefore a plane pulse reflects and refracts exactly as a time-harmonic, plane wave. When critical refraction takes place there is a frequency dependence that distorts how a pulse reflects and refracts. Moreover, this particular frequency dependence is rather interesting because it depends only on the sign of ω. To explore this further[3] we construct an incident plane-wave pulse, namely

$$u_{30}(t - \hat{\boldsymbol{p}}_0 \cdot \boldsymbol{x}/c) = \frac{1}{\pi}\Re\int_0^\infty A_0(\omega)\,e^{-i\omega(t-\hat{\boldsymbol{p}}_0\cdot\boldsymbol{x}/c)}\,d\omega, \tag{3.28}$$

[3] This argument has been taken from Friedlander (1947).

where we have used the fact that $A_0(\omega) = A_0^*(-\omega)$ because u_{30} is real. $A_0(\omega)$ is no longer simply a positive real constant, but can now be an arbitrary amplitude with a frequency dependence. From the linearity of the problem it follows that the reflected pulse is given by

$$u_{31}(t - \hat{\boldsymbol{p}}_1 \cdot \boldsymbol{x}/c) = \frac{1}{\pi} \Re \int_0^\infty A_0(\omega) e^{-i2\varphi} e^{-i\omega(t-\hat{\boldsymbol{p}}_1\cdot\boldsymbol{x}/c)}\, d\omega, \tag{3.29}$$

while the refracted pulse is given by

$$u_{32}(\boldsymbol{x}, t) = |T(\theta_0)| \frac{1}{\pi} \Re \int_0^\infty A_0(\omega) e^{-i\varphi}\, e^{-\omega x_2 \sinh\beta/\bar{c}} e^{-i\omega(t - x_1\cosh\beta/\bar{c})}\, d\omega. \tag{3.30}$$

The reflected pulse retains many of the features of the incident pulse and most importantly the argument indicating that the wave is plane. The refracted pulse is rather more complicated.

We rewrite the reflected pulse in the form

$$u_{31}(t - \hat{\boldsymbol{p}}_1 \cdot \boldsymbol{x}/c) = \cos 2\varphi\; u_{30}(t - \hat{\boldsymbol{p}}_1 \cdot \boldsymbol{x}/c) + \sin 2\varphi\; v(t - \hat{\boldsymbol{p}}_1 \cdot \boldsymbol{x}/c), \tag{3.31}$$

where

$$v(t - \hat{\boldsymbol{p}}_1 \cdot \boldsymbol{x}/c) = \frac{1}{\pi} \Im \int_0^\infty A_0(\omega)\, e^{-i\omega(t-\hat{\boldsymbol{p}}_1\cdot\boldsymbol{x}/c)} d\omega. \tag{3.32}$$

The incident pulse is reflected with a diminished amplitude and a new pulse has been excited. The function v is called the allied function (Titchmarsh, 1948).

To make this a little more concrete we examine an incident pulse of the form $u_{30}(t) = H(t) - H(t-a)$, where $H(t)$ is the Heaviside function and a measures the length of the pulse. Its allied function is

$$v(t) = \frac{1}{\pi} \ln\left|\frac{a-t}{t}\right|. \tag{3.33}$$

The reflected pulse then becomes

$$\begin{aligned} u_{31}(t - \hat{\boldsymbol{p}}_1 \cdot \boldsymbol{x}/c) &= \cos 2\varphi\; u_{30}(t - \hat{\boldsymbol{p}}_1 \cdot \boldsymbol{x}/c) \\ &\quad + \sin 2\varphi\; \frac{1}{\pi} \ln\left|\frac{t - a - \hat{\boldsymbol{p}}_1 \cdot \boldsymbol{x}/c}{t - \hat{\boldsymbol{p}}_1 \cdot \boldsymbol{x}/c}\right|. \end{aligned} \tag{3.34}$$

Note that the second pulse is present for all time and appears to arrive before the incident pulse. This is an example of a two-sided wave. It cannot exist alone and must be part of a more extended disturbance.

To explain this anomalous behavior we must consider the transmitted pulse. Carrying out the necessary integrations gives

$$u_{32}(\boldsymbol{x}, t) = |T(\theta_0)|\frac{1}{\pi}\left\{\cos\varphi\left[\tan^{-1}\left(\frac{\tau}{\lambda}\right) - \tan^{-1}\left(\frac{\tau - a}{\lambda}\right)\right]\right.$$
$$\left. + \frac{1}{2}\sin\varphi \ln\left|\frac{(\tau - a)^2 + \lambda^2}{\tau^2 + \lambda^2}\right|\right\}, \tag{3.35}$$

where

$$\tau = t - x_1 \cosh\beta/\bar{c}, \qquad \lambda = x_2 \sinh\beta/\bar{c}. \tag{3.36}$$

Clearly the refracted pulse is also two sided and present for all time.

This is not as anomalous as it might at first appear, once we recall that the upper material is faster than the lower and that the incident plane pulse has been exciting the upper material from $t \to -\infty$, at $x_1 \to -\infty$ ($\theta_0 > 0$) in Fig. 3.3. At the far left, the incident pulse excites a disturbance in the faster material. This disturbance propagates along the interface at $\bar{c}$ and fills the whole upper material, or, at least, the region near the interface. However, the main disturbance moves along the interface at $\bar{c}/\cosh\beta = c/\sin\theta_0$. Therefore the precursor in the upper material gets ahead of the trace of the wavefront, of the incident pulse, at the interface and must somehow satisfy the boundary condition. It does so by shedding a reflected pulse into the slower material that appears before the incident one arrives, the second term in (3.34). Moreover, no energy is permanently carried into the upper material. The refracted pulse continually reradiates into the slower, lower material. It is also interesting to note that, as the distance from the interface increases, u_{32} decays as $1/x_2$ rather than as $e^{-\bar{k}x_2 \sinh\beta}$, so that the pulse decays algebraically, while the time-harmonic disturbance decays exponentially.

The origin of this phenomenon is the $\cos\theta_2$ in the expressions for $C_{\mp}$, (3.23). But $\cos\theta_2 = [1-(\bar{c}/c)^2 \sin^2\theta_0]^{1/2}$. If we regard $(\omega/c)\sin\theta_0$ as a complex variable z, then in fact the phenomenon arises from the definition of the branch of the function $(\bar{k}^2 - z^2)^{1/2} (= \bar{k}\cos\theta_2)$ that appears in the reflection and transmission coefficients. Critical refraction, and reflection, therefore, manifest themselves as branch cuts in the complex plane.

Problem 3.5 Reflection of an Acoustic Wave from a Plate

Consider the reflection of a time harmonic, plane acoustic wave from a thin elastic plate. We assume that the fluid has no viscosity so that the incident acoustic wave excites predominantly flexural motions in the thin plate. Moreover, we assume that the fluid below the plate is sufficiently dense that the elastic

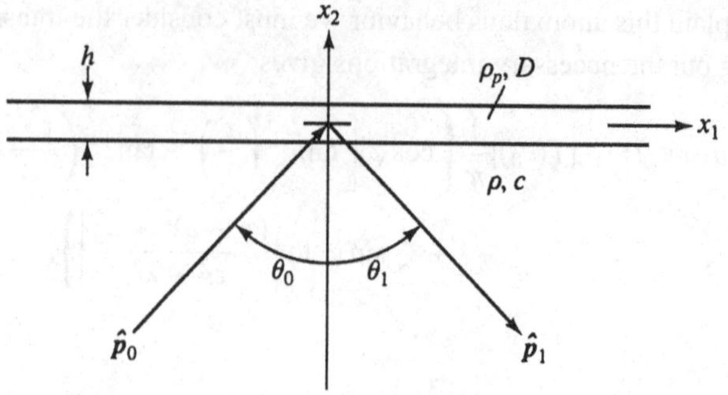

Fig. 3.5. A thin elastic plate separates a dense fluid such as water from a tenuous one such as air, which we model as a vacuum. A time harmonic, plane acoustic wave, incident from the fluid, strikes the plate and is reflected.

plate and the fluid can interact, but that the fluid above the plate is sufficiently tenuous that it can be treated as a vacuum. Water and air, respectively, satisfy these assumptions. Figure 3.5 indicates the geometry of the problem.

In the fluid

$$\nabla^2\varphi + k^2\varphi = 0 \tag{3.37}$$

where φ is the *velocity* potential, $k = \omega/c$, and a time dependence of the form $e^{-i\omega t}$ is assumed but suppressed. The pressure $p = i\omega\rho\varphi$ and the particle *displacement* is $-i\omega\boldsymbol{u} = \nabla\varphi$. The density of the fluid is ρ.

The one dimensional, flexural motion of the plate is described by

$$(\partial_1\partial_1)^2 w - k_p^4 w = p(x_1, 0)/D, \tag{3.38}$$

where $k_p^4 = \omega^2\rho_p h/D$. The terms ρ_p, h, and D are the density per unit area, the thickness, and the elastic constant for the plate, respectively. At $x_2 = 0$, the particle displacement is continuous so that $-i\omega w(x_1, x_3) = \partial\varphi/\partial x_2$. Note that phase matching must take place so that the x_1 dependence of w is determined by the incident wave. Equation (3.38) then gives a boundary condition connecting p and φ at $x_2 = 0$.

The incident and reflected waves are $\varphi_0 = A_0 e^{ik\hat{p}_0\cdot x}$ and $\varphi_1 = A_1 e^{ik\hat{p}_1\cdot x}$, respectively. Find the reflection coefficient $R(\theta_0)$. Note that $|R(\theta_0)| = 1$ so that $R(\theta_0) = -e^{i2\alpha}$. Find α. Speculate about the form of a reflected pulse.

3.4 The Rayleigh Wave

We have seen that critical refraction produces a wave in the faster material that clings to the interface. However, this wave could not exist without being continuously sustained by a wave from the interior. In contrast there are waves that cling to a surface or interface that are self-sustaining. Their particle displacement decays exponentially with distance away from the surface or interface, in the time-harmonic approximation. Such waves are also referred to as surface or interfacial waves. It is one of the distinctive features of linear elasticity that a traction-free surface guides such a wave, called a Rayleigh wave. Both here and in Chapter 5, it is demonstrated that this latter surface wave, and ones similar to it, arises from a pole rather than a branch point.

3.4.1 The Time-Harmonic Wave

We consider an elastic half-space with the x_1 coordinate lying along the surface and the positive x_2 coordinate pointing into the interior (the reverse of that shown in Figs. 3.1 and 3.3). We seek a time-harmonic wave whose particle displacement is inplane and that decays in the positive x_2 direction. We use a scalar potential φ and a single component ψ_3 of the vector potential [the divergence condition (1.19) is then automatically satisfied], and assume that

$$\varphi = Ce^{-\beta\gamma_L x_2} e^{i\beta x_1}, \qquad \psi_3 = De^{-\beta\gamma_T x_2} e^{i\beta x_1}, \tag{3.39}$$

where $\beta = \omega/c$ and c is the unknown wavespeed along the surface $x_2 = 0$. To satisfy (1.21) and (1.22) for the potentials,

$$\gamma_L = \left(1 - \frac{c^2}{c_L^2}\right)^{1/2}, \qquad \gamma_T = \left(1 - \frac{c^2}{c_T^2}\right)^{1/2}. \tag{3.40}$$

The boundary conditions at $x_2 = 0$ are $\tau_{22} = \tau_{21} = 0$. To satisfy these boundary conditions,

$$\begin{bmatrix} 2 - c^2/c_T^2 & 2i\gamma_T \\ -2i\gamma_L & 2 - c^2/c_T^2 \end{bmatrix} \begin{bmatrix} C \\ D \end{bmatrix} = \begin{bmatrix} 0 \\ 0 \end{bmatrix}. \tag{3.41}$$

Setting the determinant of the matrix to zero, the condition for a nontrivial solution, gives

$$[2 - (c^2/c_T^2)]^2 - 4\gamma_L\gamma_T = 0. \tag{3.42}$$

This equation, or an insignificant modification of it, is called the Rayleigh equation. We examine it momentarily. It has a real positive solution $c = c_r$, giving the wavespeed of the surface wave. Note that γ_L and γ_T must be real if we are to have decay into the interior; thus $0 < c_r < c_T$. Returning to (3.41) we find that

$$\frac{C}{D} = \frac{-2i\gamma_T}{(2 - c_r^2/c_T^2)}. \tag{3.43}$$

With the use of (1.19), the particle displacement components are

$$u_1 = i\beta_r \frac{C}{2}\left[2e^{-\beta_r\gamma_L x_2} - \left(2 - \frac{c_r^2}{c_T^2}\right)e^{-\beta_r\gamma_T x_2}\right]e^{i\beta_r x_1},$$

$$u_2 = \frac{-\beta_r C}{2\gamma_T}\left[2\gamma_L\gamma_T\, e^{-\beta_r\gamma_L x_2} - \left(2 - \frac{c_r^2}{c_T^2}\right)e^{-\beta_r\gamma_T x_2}\right]e^{i\beta_r x_1}, \tag{3.44}$$

where $\beta_r = \omega/c_r$. This then is the form of the time harmonic, Rayleigh wave.

3.4.2 Transient Wave

Previously we have seen that a critically refracted wave becomes a two-sided wave in the time domain. Much the same thing happens when a time-harmonic Rayleigh wave is mapped into the time domain. In this case the frequency dependence enters through the terms $e^{-\beta_r\gamma_L x_2}$ and $e^{-\beta_r\gamma_T x_2}$. Proceeding as we did in (3.28)–(3.30), the transient displacement components are given by

$$u_1(t - x_1/c_r, x_2) = \frac{1}{\pi}\Re\int_0^\infty A_0(\omega)\left[2e^{-(\omega/c_r)x_2\gamma_L} - \left(2 - \frac{c_r^2}{c_T^2}\right)e^{-(\omega/c_r)x_2\gamma_T}\right]e^{-i\omega(t - x_1/c_r)}\,d\omega,$$

$$u_2(t - x_1/c_r, x_2) = \frac{1}{\pi}\Re\int_0^\infty A_0(\omega)e^{i\pi/2}\gamma_T^{-1}\left[2\gamma_L\gamma_T\, e^{-(\omega/c_r)x_2\gamma_L} - \left(2 - \frac{c_r^2}{c_T^2}\right)e^{-(\omega/c_r)x_2\gamma_T}\right]e^{-i\omega(t - x_1/c_r)}\,d\omega, \tag{3.45}$$

where $A_0(\omega)$ has the same form as in (3.28)–(3.30) and would be set by the source. For simplicity we take $A_0(\omega) = \pi C$, where C is a constant. Note the $e^{i\pi/2}$ in u_2. Performing the integrations gives

$$u_1(t - x_1/c_r, x_2) = \frac{2C(x_2/c_r)\gamma_L}{[(x_2/c_r)^2\gamma_L^2 + (t - x_1/c_r)^2]} - \frac{C[2 - (c_r^2/c_T^2)][(x_2/c_r)\gamma_T]}{[(x_2/c_r)^2\gamma_T^2 + (t - x_1/c_r)^2]},$$

$$u_2(t - x_1/c_r, x_2) = -\frac{2C(t - x_1/c_r)\gamma_L}{[(x_2/c_r)^2\gamma_L^2 + (t - x_1/c_r)^2]} + \frac{C}{\gamma_T}\frac{[2 - (c_r^2/c_T^2)](t - x_1/c_r)}{[(x_2/c_r)^2\gamma_T^2 + (t - x_1/c_r)^2]}. \tag{3.46}$$

Because of our choice of A_0, these pulses are more singular than real ones, but nevertheless they exhibit the basic features of interest. Note that there is no leading wavefront. We have a two-sided wave. Using our approximations, we have implicitly assumed that the leading wavefront, which is that of a compressional pulse, has already propagated everywhere along the surface. The compressional pulse was excited at $t \to -\infty$ and propagates outward at wavespeed c_L, where $c_L > c_T > c_r$. Also note that the exponential decay of the time-harmonic disturbance has become an algebraic decay. Lastly, the reader, having noted the presence of the $e^{i\pi/2}$ in (3.45), may find it interesting to show that the u_2 component is essentially the allied function of the u_1 component.

3.4.3 The Rayleigh Function

We are thus left with examining the roots of (3.42). The most important feature of this equation is that the function on the left-hand side contains two radicals whose branches must be defined with care. To do so it is more convenient to work with $R(s)$, a slightly modified version of (3.42). This function, sometimes called the Rayleigh function, is given by

$$R(s) = \left(2s^2 - s_T^2\right)^2 - 4s^2\left(s^2 - s_L^2\right)^{1/2}\left(s^2 - s_T^2\right)^{1/2}, \qquad (3.47)$$

where $s = 1/c$ and $s_I = 1/c_I$.

Recall that the denominator of the reflection coefficient for the inplane reflection is $A_+(\theta_0)$, given in (3.12). Setting $s = s_L \sin\theta_0 = s_T \sin\theta_2$, we find that $A_+(\theta_0) = R(s_L \sin\theta_0)/(s_L s_T)^2$. This feature is not accidental. *A pole in the reflection coefficient implies that a surface wave exists.* The question then is whether the pole is real or complex, and, if complex, is its imaginary part such that the surface wave is physically meaningful. The slownesses are ordered as $s > s_T > s_L$ for a surface wave to exist, so that $\sin\theta_0 = s/s_L > 1$, and thus θ_0 must be complex. A homogeneous plane wave striking a traction-free boundary cannot excite a surface wave.

Problem 3.6 A Surface Wave Supported by an Impedance

Consider a time harmonic disturbance $u(x_1, x_2)$ that satisfies

$$\partial_1\partial_1 u + \partial_2\partial_2 u + k^2 u = 0, \qquad (3.48)$$

where $k = \omega/c$. Further, consider the same general geometry as shown in Fig. 3.5, but with the plate replaced by an impedance. That is, the boundary condition at $x_2 = 0$ becomes

$$\partial_2 u = ikZu, \qquad Z = R + iX, \tag{3.49}$$

with Z the impedance. The impedance may depend upon ω, but not upon the angle of incidence of the wave. That is, the response at a given point on the boundary does not depend upon the response of the surrounding points. Taking as the incident wave u_0 one identical to (3.17), and as the reflected wave u_1 one identical to (3.18), calculate the reflection coefficient.

Next show that the surface wave

$$u = Ce^{-b|x_2|}e^{iax} \tag{3.50}$$

is also a solution. a and b are positive and real. C is a constant. What restrictions must R and X satisfy for a surface wave to exist? Find a and b in terms of a Z that permits a surface wave to exist. For this same Z show that the reflection coefficient has a pole at a particular θ_0, say θ_{0s}. Using this θ_{0s}, show that the reflected wave takes the form of the surface wave, (3.50).

3.4.4 Branch Cuts

The Rayleigh function contains two radicals of the form

$$\gamma = (\alpha^2 - k^2)^{1/2}, \tag{3.51}$$

where α is the independent variable and k is known. Almost always, k is a wavenumber. Thus we may, by imagining that the material in which the wave is propagating is slightly lossy, set $k = k_r + ik_i$, where $|k_i/k_r| \ll 1$. By considering a plane wave propagating in the positive x direction with wavenumber k, namely, $e^{ik_r x}e^{-k_i x}$, we see that k_r and k_i must both be positive. Looking back at (3.39) and (3.40), we note that for a surface wave $\Re(\gamma_L) \geq 0$ and $\Re(\gamma_T) \geq 0$. Accordingly, we want to define the branches of the radical, so that $\Re(\gamma) \geq 0 \,\forall\, \alpha$. The curves $\Re(\gamma) = 0$ therefore define the branch cuts.

It is hard to work with γ directly, but somewhat easier to begin by working[4] with γ^2. Setting $\alpha = \sigma + i\tau$, we can write γ^2 as

$$\gamma^2 = \left[(\sigma^2 - \tau^2) - (k_r^2 - k_i^2)\right] + 2i(\sigma\tau - k_r k_i). \tag{3.52}$$

We next partition the α plane, as shown in Fig 3.6, using the curves $\Re(\gamma^2) = 0$ and $\Im(\gamma^2) = 0$. Perturbing α slightly, we find that $\Re(\gamma^2) < 0$ between the

[4] I first learned of this argument from Mittra and Lee (1971).

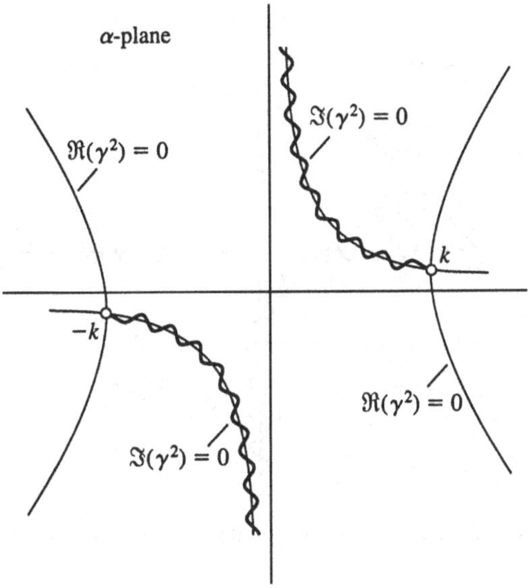

Fig. 3.6. The complex α plane with a sketch of the curves $\Re(\gamma^2) = 0$ and $\Im(\gamma^2) = 0$. From these curves we select the branch cuts indicated by the wavy overlying line on the curves $\Im(\gamma^2) = 0$.

two curves $\Re(\gamma^2) = 0$, while $\Im(\gamma^2) < 0$ between the two curves $\Im(\gamma^2) = 0$. Expressing γ as $|\gamma|e^{i\theta}$, we note that, if $\Re(\gamma) > 0$, then $|\theta| < \pi/2$ and thus $|2\theta| < \pi$. The branch cuts, $\Re(\gamma) = 0$, therefore must lie along the curves $2\theta = \pm\pi$. That is along $\Im(\gamma^2) = 0$, where $\Re(\gamma^2) < 0$. These are the curves in Fig. 3.6 overlain with the wavy line. In the limit as $k_i \to 0$, we arrive at the branch cuts shown in Fig. 3.7.

To investigate this branch, express γ as

$$\gamma = (r_1 r_2)^{1/2}\left[\cos\left(\frac{\theta_1 + \theta_2}{2}\right) + i \sin\left(\frac{\theta_1 + \theta_2}{2}\right)\right], \tag{3.53}$$

where

$$(\alpha - k)^{1/2} = r_1^{1/2} e^{i\theta_1/2}, \qquad (\alpha + k)^{1/2} = r_2^{1/2} e^{i\theta_2/2}. \tag{3.54}$$

Figure 3.7 indicates from what reference the angles are measured. Note that $\Re(\gamma) \geq 0 \,\forall\, \alpha$, while $\Im(\gamma) > 0$ in the first and third quadrants and negative in the other two. As well, note that to go from the first to the second quadrant, and remain on the same Riemann sheet, we must first move into the fourth and then poke through the gap at the origin into the second.

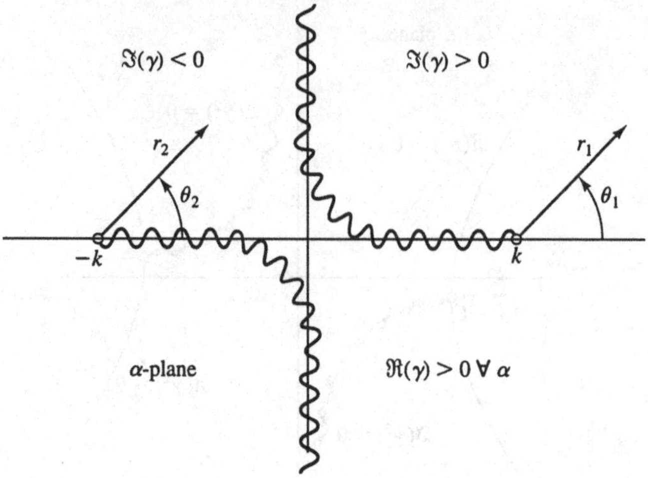

Fig. 3.7. The complex α plane as $k_i \to 0$. $\Re(\gamma) \geq 0 \,\forall\, \alpha$ on the Riemann sheet exhibited in the drawing.

Returning to the Rayleigh function $R(s)$, (3.47), we note that there are two radicals appearing as the product $(s^2 - s_L^2)^{1/2}(s^2 - s_T^2)^{1/2}$. Both radicals are defined as just indicated. However, the product is only discontinuous across a branch cut connecting s_L and s_T, and one connecting $-s_L$ and $-s_T$. These are the appropriate branch cuts for the product. Hence the Rayleigh function $R(s)$ is defined as that branch occupying the Riemann sheet $\Re(s\gamma_L) \geq 0$, $\Re(s\gamma_T) \geq 0 \,\forall\, s$ and cut as just indicated. Recall that $s = 1/c$. We next seek a root or roots to $R(s) = 0$ on this Riemann sheet.

First, note that if s_r is a root then so is $-s_r$. Second, at $s = s_L$ and at $s = s_T$, $R(s) > 0$ but as $s \to \infty$ along the positive real s axis, $R(s)$ becomes negative. Thus a real root s_r, with $s_T < s_r < \infty$, exists. By symmetry a root $-s_r$ also exists. Third, does $R(s)$ have any other roots *on this Riemann sheet*? The answer is no. This can be proven conclusively by using a theorem sometimes called the *argument principle* (Ablowitz and Fokas, 1997). This theorem allows one to systematically calculate the number of poles and zeros of a function by performing a contour integral of its logarithmic derivative. Achenbach (1973) does this in some detail, though the original argument may be found in Cagniard (1962). Lastly, do the roots on the other Reimann sheets ever manifest themselves? Yes, they do when they lie close to a branch cut. This point is discussed in Aki and Richards (1980) in their description of the reflection of spherical waves from a traction-free surface.

While knowing that only the Rayleigh slownesses $\pm s_r$ exist on the Reimann sheet of interest is important, actually solving for s_r is less so. It can be calculated to good accuracy by using the formula

$$\kappa_r = (0.86 + 1.12\nu)/(1 + \nu) \tag{3.55}$$

The term $\kappa_r = s_T/s_r$ and ν is Poisson's ratio. For many materials κ_r is close to 0.9.

The branches of (2.23) and (2.31) were selected as just indicated. *However,* note that $k_1 = i\gamma$ in (2.23) and $k_3 = i\gamma$ in (2.31). In particular, in the case of (2.23), the complex k_2 plane is cut so that $\Im(k_1) > 0 \,\forall\, k_1$, corresponding exactly with the branch of γ just selected. Further, the contour in (2.24) starts at $-\infty$ in the second quardant, passes above the branch cut, pokes into the fourth quadrant, passes below the branch cut, and ends at $+\infty$. The radicals we encounter will often present themselves in the form $\xi = (k^2 - \alpha^2)^{1/2}$, rather than as that in (3.51). The two are connected by $\xi = i\gamma$. Only when considering the Cagniard–deHoop integration technique in Section 5.1 do we select a different branch.

References

Ablowitz, M.J. and Fokas, A.S. 1997. *Complex Variables,* pp. 259–265. New York: Cambridge.

Achenbach, J.D. 1973. *Wave Propagation in Elastic Solids,* pp. 189–191. Amsterdam: North-Holland.

Aki, K. and Richards, P.G. 1980. *Quantitative Seismology, Theory and Methods,* Vol. 1, pp. 319–333. San Francisco: Freeman.

Auld, B.A. 1990. *Acoustic Fields and Waves in Solids,* 2nd ed., Vol. 2, pp. 1–57. Malabar, FL: Krieger.

Cagniard, I. 1962. *Reflection and Refraction of Progressive Seismic Waves,* pp. 42–49. Translated and revised by E.A. Flinn and C.H. Dix. New York: McGraw-Hill.

Friedlander, F.G. 1974. On the total reflection of plane waves. *Quart. J. Mech. Appl. Math.* **1**: 379–383.

Mittra, R. and Lee, S.W. 1971. *Analytical Technique in the Theory of Guided Waves,* pp. 20–23. New York: Macmillan.

Titchmarsh, E.C. 1948. *Introduction to the Theory of Fourier Integrals,* 2nd ed., pp. 119–121, 147–148, and elsewhere. Oxford: Clarendon Press.

4

Green's Tensor and Integral Representations

Synopsis

In Chapter 4 we discuss the formulation of integral representations of solutions to rather general problems in elastic-wave propagation. Two constructions are used: the reciprocity identity and the Green's tensor for a full space. For these representations for an infinite domain to be derived, the principle of limiting absorption is introduced. This is needed for time-harmonic problems because the disturbance, in a sense, has been going on forever, resulting in no initial wavefront being present. Moreover, we establish a uniqueness result, indicating as we do so both the role of the principle of limiting absorption and that of specifying an edge condition. The chapter closes with an example that uses these ideas to develop an integral representation for the scattering of an acoustic wave by an elastic inclusion.

4.1 Introduction

In Chapter 2 we moved away from discussing plane waves to an introduction of plane-wave spectral representations in Section 2.3. This allowed us to discuss more general wavefields and to understand their propagation characteristics in terms of those of plane waves. We continue with this general theme, but construct, in this chapter, both far more general representations and ones in physical space rather than in wavenumber space. Though we make limited use of it in the chapters that follow, this material is very important because it is the basis for formulating elastic-wave problems in a form suitable to be analyzed numerically. A general survey of many useful representation results is provided by deHoop (1995).

When the wavefield is time dependent, it is straightforward (straightforward is not a synonym for easy) to seek its representations (Friedlander, 1958; Achenbach, 1973; Hudson, 1980). We must work in a four-dimensional space, the fourth dimension being time. Nevertheless, the wavefield propagates outward from its sources at a finite speed and thus for a finite time it occupies a region that is bounded. However, a representation in this form is less useful than it might seem because both the source and receiver of the wavefield have their own frequency behaviors, and to incorporate those effects in an overall propagation model the Fourier component of the wavefield rather than, say, its time-dependent response to a sudden impulse is of greater use. Thus we consider almost exclusively time-harmonic wavefields.

When working with a time-harmonic wavefield, though we need only work with regions in three-dimensional space, we have imagined that the wavefield was excited at the time minus infinity, and it has thus, even for a finite positive time, filled all of space. In constructing representations for various wavefields, we must *unambiguously* identify waves that are outgoing from their source of excitation. Moreover, at some stage we must send the outer surfaces, over which integrals are being taken, to infinity and must be assured that these integrals either go to zero or a finite, unambiguously identified value. In this chapter we use the principle of limiting absorption as the technical device to achieve these outcomes.

4.2 Reciprocity

Proposition 4.1. *In a bounded region $\mathcal{R}_x$ with surface $\partial\mathcal{R}_x$, the equations of motion in the time-harmonic approximation for two reciprocating elastic wavefields, indicated by the superscripts* 1 *and* 2, *are*

$$\partial_j \tau_{ji}^{1,2} + \rho f_i^{1,2} + \rho\omega^2 u_i^{1,2} = 0. \tag{4.1}$$

The tractions on $\partial\mathcal{R}_x$ are $t_i^{1,2} = \tau_{ji}^{1,2}\hat{n}_j$. The unit normal $\hat{\boldsymbol{n}}$ to $\partial\mathcal{R}_x$ is outward. Then

$$\int_{\mathcal{R}_x} \left(f_i^2 u_i^1 - f_i^1 u_i^2 \right) \rho \, dV = \int_{\partial\mathcal{R}_x} \left(u_i^2 t_i^1 - u_i^1 t_i^2 \right) dS. \tag{4.2}$$

Proof. We take the scalar product of each equation in (4.1) with the particle displacement of its reciprocating wavefield and subtract. This gives

$$\rho \left(f_i^2 u_i^1 - f_i^1 u_i^2 \right) = -u_i^1 \partial_j \tau_{ji}^2 + u_i^2 \partial_j \tau_{ji}^1. \tag{4.3}$$

Using the relation between stress and strain, (1.3), we note that $\tau_{ji}^1 \partial_j u_i^2 = \tau_{ji}^2 \partial_j u_i^1$ so that the right-hand side of (4.3) equals $\partial_j(\tau_{ij}^1 u_i^1 - \tau_{ij}^2 u_i^1)$; (4.2) follows after integration. □

There is one special case of (4.2) of interest. Let us assume that the integral over ∂R_x vanishes and that $f_i^{1,2} = a_i^{1,2}\delta(\boldsymbol{x} - \boldsymbol{x}^{1,2})$, where the $a_i^{1,2}$ are constant vectors. Then the reciprocity statement becomes $a_i^2 u_i^1(\boldsymbol{x}^2) = a_i^1 u_i^2(\boldsymbol{x}^1)$. This is the origin of the careless statement "the source and receiver can be interchanged by reciprocity."

4.3 Green's Tensor

In the proposition to follow we shall use the three-dimensional Fourier transform pair given by

$$^{*}u(\boldsymbol{k}) = \int_{-\infty}^{\infty} u(\boldsymbol{x}) e^{-i\boldsymbol{k}\cdot\boldsymbol{x}} d\boldsymbol{x}, \tag{4.4}$$

$$u(\boldsymbol{x}) = \frac{1}{(2\pi)^3} \int_{-\infty}^{\infty} {}^{*}u(\boldsymbol{k}) e^{i\boldsymbol{k}\cdot\boldsymbol{x}} d\boldsymbol{k}. \tag{4.5}$$

This is a straightforward generalization of (1.43) and (1.44). The differential in physical space $d\boldsymbol{x} = dx_1\, dx_2\, dx_3$ and that in the transform space $d\boldsymbol{k} = dk_1\, dk_2\, dk_3$.

Proposition 4.2. *The particle displacement u_i^G is the solution to*

$$(\lambda + \mu)\partial_i \partial_k u_k^G + \mu \partial_j \partial_j u_i^G + \rho\omega^2 u_i^G = -\rho \hat{a}_i \delta(\boldsymbol{x}), \tag{4.6}$$

subject to the condition that, as $x \to \infty$, the solution represent an outgoing wavefield. $|\boldsymbol{x}| = x$. Here $\hat{\mathbf{a}}$ is a constant unit vector giving the direction of the point force at the origin. The Green's tensor[1] for a full space, u_{ik}^G, is related to u_i^G by $u_i^G = \hat{a}_k u_{ik}^G$, with u_{ik}^G given by

$$u_{ik}^G = \frac{1}{k_T^2 c_T^2} \{\partial_i \partial_k [-G(k_L x) + G(k_T x)] + k_T^2 \delta_{ik} G(k_T x)\}, \tag{4.7}$$

where

$$G(k_I x) = (1/4\pi x) e^{ik_I x}, \qquad I = L, T. \tag{4.8}$$

Note that we may replace $\boldsymbol{x}$ with the more general vector $\boldsymbol{x} - \boldsymbol{x}'$, where $\boldsymbol{x}'$ is the position vector of the point of action of the delta function.

[1] It is possibly better to describe the pair (u_{ik}^G, τ_{ijk}^G) as a Green's state.

Proof.[2] Fourier transforming (4.6) in all three spatial variables and solving the resulting algebraic equations give

$$^*u_i^G = -\frac{1}{c_T^2}\frac{\left(k_L^2 - k^2\right)\hat{a}_i + \left(1 - c_T^2/c_L^2\right)k_i k_k \hat{a}_k}{\left(k_L^2 - k^2\right)\left(k_T^2 - k^2\right)}, \tag{4.9}$$

where $k^2 = \boldsymbol{k}\cdot\boldsymbol{k}$. From the definition of u_{ij}^G, and with some rewriting, we arrive at

$$^*u_{ik}^G = \frac{1}{c_T^2}\left[\frac{k_i k_k}{k_T^2}\left(\frac{1}{k^2 - k_L^2} - \frac{1}{k^2 - k_T^2}\right) + \frac{\delta_{ik}}{k^2 - k_T^2}\right]. \tag{4.10}$$

Note that ∂_i in physical space transforms to ik_i in transform space and vice versa. Therefore, to transform (4.10) to physical space, all that we have to do is evaluate integrals of the form

$$I_I = \frac{1}{(2\pi)^3}\int_{-\infty}^{\infty}\frac{e^{i\boldsymbol{k}\cdot\boldsymbol{x}}}{k^2 - k_I^2}\,d\boldsymbol{k}. \tag{4.11}$$

To evaluate (4.11) we introduce, in transform space, a spherical coordinate system (k, ξ, ν), where the azimuthal axis is chosen in the direction $\boldsymbol{x}$ and ξ is the azimuthal angle. Note that this is identical, apart from the change in symbols, to the coordinate transformation (2.33) and (2.37) used to construct an angular spectrum representation of a spherical wave. The transformation is defined as

$$k_1 = k\sin\xi\,\cos\nu, \quad k_2 = k\sin\xi\,\sin\nu, \quad k_3 = k\cos\xi. \tag{4.12}$$

The integral (4.11) can then be expressed as

$$I_I = \frac{1}{(2\pi)^3}\int_0^{\infty}\int_0^{\pi}\int_0^{2\pi}\frac{k^2\sin\xi}{k^2 - k_I^2}e^{ikx\cos\xi}\,dk\,d\xi\,d\nu. \tag{4.13}$$

The integrations over the angles are readily done, and we are left with evaluating

$$I_I = \frac{-i}{(2\pi)^2 x}\int_{-\infty}^{\infty}\frac{ke^{ikx}}{k^2 - k_I^2}dk. \tag{4.14}$$

Recall that $e^{i(k_I x - \omega t)}$ is an outgoing wave. Thus to satisfy the condition that waves be outgoing as $x \to \infty$, the contour of the integral must pass above the

[2] Contrast how we have used a three-dimensional transform in this problem, but have used a two-dimensional one for the very similar problem of constructing a plane-wave representation of a spherical wave in (2.36).

pole at $-k_I$ but below that at k_I. Then closing the contour in the upper half of the k plane, where the integral is convergent, gives an outgoing wave and hence gives (4.8) with $G(k_I x) = I_I$. Thus (4.7) follows. □

The corresponding Green's stress τ^G_{ijk} is calculated by using

$$\tau^G_{ijk} = c_{ijlm}\partial_l u^G_{mk}, \tag{4.15}$$

where

$$c_{ijlm} = \lambda\delta_{ij}\delta_{lm} + \mu(\delta_{il}\delta_{jm} + \delta_{im}\delta_{jl}). \tag{4.16}$$

These last two equations are simply restatements of (1.3).

Problem 4.1 Two-Dimensional Green's Function

In this problem we ask the reader to calculate the time harmonic, antiplane shear Green's function, u_3^G. In fact it will turn out to be identical to the angular spectrum representation of a cylindrical wave, cited in Section 2.3.3. The equation to be solved[3] is

$$\partial_\alpha\partial_\alpha u_3^G + k^2 u_3^G = -\delta(\boldsymbol{x}), \tag{4.17}$$

where $\boldsymbol{x}$ has components (x_1, x_2). At infinity u_3^G represents an outgoing wave. Show that u_3^G can be expressed as

$$u_3^G(\boldsymbol{x}) = \frac{i}{4\pi}\int_{-\infty}^{\infty} e^{i(k_1 x_1 + k_2|x_2|)}\frac{dk_1}{k_2}. \tag{4.18}$$

How are the branches of the radical k_2 defined? Next introduce the transformations, $k_1 = k_T\cos\xi$, $x_1 = \rho\cos\theta$, $|x_2| = \rho\sin\theta$, and show that (4.18) reduces to

$$u_3^G(\rho) = -\frac{i}{4\pi}\int_C e^{ik_T\rho\cos(\theta-\xi)}\,d\xi. \tag{4.19}$$

Most importantly show that C is a contour beginning near $\pi - i\infty$ and ending near $i\infty$. The integral representation of the Hankel function $H_0^{(1)}(k\rho)$ (Sommerfeld, 1964b) is such that $u_3^G = (i/4)H_0^{(1)}(k\rho)$.

[3] The right-hand side of this equation should be multiplied by c^{-2}, if the reader wants its dimensional form to be identical to (4.6).

4.3.1 Notes

1. For future work we shall need the farfield results. The farfield is that region for which $|k_I \boldsymbol{x}'| \ll |k_I \boldsymbol{x}|$ and $|k_I \boldsymbol{x}| \gg 1$. Approximating $|\boldsymbol{x} - \boldsymbol{x}'|$ by using the law of cosines gives

$$|\boldsymbol{x} - \boldsymbol{x}'| \sim x - (\hat{\boldsymbol{p}} \cdot \boldsymbol{x}'), \qquad \hat{\boldsymbol{p}} = \boldsymbol{x}/x. \tag{4.20}$$

With this approximation, (4.7), with x replaced by $|\boldsymbol{x} - \boldsymbol{x}'|$, splits into two parts, namely

$$u_{ik}^{G} \sim u_{ik}^{GL} + u_{ik}^{GT}, \tag{4.21}$$

where

$$u_{ik}^{GL} = \frac{1}{c_L^2} \hat{p}_i \hat{p}_k \frac{e^{ik_L x}}{4\pi x} e^{-ik_L \hat{\boldsymbol{p}} \cdot \boldsymbol{x}'}, \tag{4.22}$$

$$u_{ik}^{GT} = \frac{1}{c_T^2} (\delta_{ik} - \hat{p}_i \hat{p}_k) \frac{e^{ik_T x}}{4\pi x} e^{-ik_T \hat{\boldsymbol{p}} \cdot \boldsymbol{x}'}. \tag{4.23}$$

Further, using (4.15) and (4.16), we have

$$\tau_{jik}^{G} \hat{p}_j \sim \tau_{jik}^{GL} \hat{p}_j + \tau_{jik}^{GT} \hat{p}_j, \tag{4.24}$$

where

$$\tau_{jik}^{GL} \hat{p}_j = ik_L(\lambda + 2\mu) u_{ik}^{GL}, \tag{4.25}$$

$$\tau_{jik}^{GT} \hat{p}_j = ik_T \mu u_{ik}^{GT}. \tag{4.26}$$

Each of the above expansions is carried out only to $\mathrm{O}[(k_I x)^{-1}]$ in the amplitude while the additional term $(ik_I \hat{\boldsymbol{p}} \cdot \boldsymbol{x}')$ is retained in the exponential terms. The character of a wavefield is more sharply determined by its phase than by its amplitude. As before, $I = L$ or T.

2. There are a number of very localized sources that can be constructed from point forces and their derivatives (Hudson, 1980). Here we consider one such construction, the *center of compression.* Consider a special (three-dimensional) body force, namely

$$\boldsymbol{f} = -c_L^2 F(t) \partial_x \left[\frac{\delta(x)}{4\pi x^2} \right] \hat{\boldsymbol{p}}, \tag{4.27}$$

where $\boldsymbol{x}'$ is set to zero, $|\boldsymbol{x}| = x$, and $\hat{\boldsymbol{p}}$ is defined in (4.20).

The scalar potential φ satisfies

$$\frac{1}{x^2}\partial_x(x^2\partial_x\varphi) - \frac{1}{c_L^2}\partial_t\partial_t\varphi = F(t)\frac{\delta(x)}{4\pi x^2} \tag{4.28}$$

and the vector potential $\psi = 0$. This particular source is useful because it excites only compressional waves. The solution to (4.28), satisfying the condition that the wave be outgoing from its source, is

$$\varphi = -\frac{1}{4\pi x}F\left(t - \frac{x}{c_L}\right). \tag{4.29}$$

The center of compression can be thought of as a small uniformly pressurized cavity or as three dipoles without moment.

Problem 4.2 A Causal Green's Function

Problem 1. Show that (4.29) is the solution to (4.28), subject to the condition that the wave be outgoing from the source. Here (4.29) is the free-space causal Green's function for (4.28) if $F(t) = \delta(t)$. Begin by showing that $x\varphi = f(t - x/c_L)$ is a solution, for $x \neq 0$. Then integrate the equation of motion, (4.28) over a small sphere of radius ϵ, centered at the origin, and, taking the limit as $\epsilon \to 0$, identify $f(t)$.

Problem 2. Equation (4.27) is a three-dimensional body force. In two dimensions, (x_1, x_2), the term in brackets becomes $\delta(x)/(2\pi x)$. Find the response φ for a two-dimensional center of compression or line of compression.

4.4 Principle of Limiting Absorption

Figure 4.1 shows a typical region $\mathcal{R}_x$ within which we shall work. It is a large spherical region with radius x. Both the bounded subregions contained within $\mathcal{R}_x$, $\mathcal{S}$ and $\mathcal{R}$, will be sources of waves. Two questions arise. (1) How do we determine unambiguously which waves are outgoing from these sources? (2) In the arguments that follow we shall ask that the integrals over $\partial\mathcal{R}_x$ vanish or approach some finite known value as $x \to \infty$. How do we ensure that this happens? As we indicated in Section 4.1, this issue does not arise in the time-dependent case because any outgoing wavefield will propagate toward infinity at a finite speed and, therefore, for a finite t the surface $\partial\mathcal{R}_x$, as $x \to \infty$, will eventually pass into a region that the wavefield has not reached. However, in

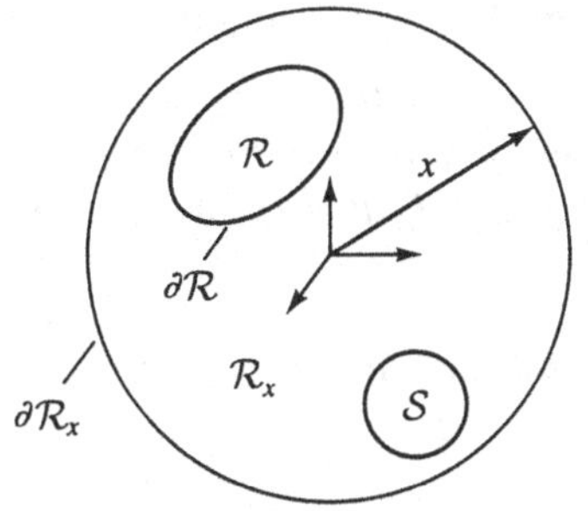

Fig. 4.1. The large spherical region $\mathcal{R}_x$, with surface $\partial\mathcal{R}_x$, has a radius $|\boldsymbol{x}| = x$. In several of the arguments $x \to \infty$. Within are two bounded subregions, $\mathcal{S}$ and $\mathcal{R}$. The first contains sources that excite an incident wavefield, while the second is a scatterer, with surface $\partial\mathcal{R}$. The unit normal $\hat{\boldsymbol{n}}$ points out from $\mathcal{R}_x \backslash \mathcal{R}$.

the time-harmonic case, we are imagining that the wavefield started abruptly at $t = -\infty$ and thus it now fills all space. Stated somewhat differently, there is no initial wavefront because we have imposed no initial conditions.

We solve this problem by demanding that $\omega = \omega_0 + i\epsilon$, where ω_0 and $\epsilon > 0$ are real. Hence, $\lim_{t\to-\infty} f(\boldsymbol{x})e^{i(kx-\omega t)} = 0$, for x fixed, so that the wavefield is forced to zero. This gives the missing initial conditions. Though, as $t \to \infty$ the wavefield becomes unbounded, t will always be taken as finite and, in a time-harmonic problem, often never appears explicitly in the calculation. It in no sense indicates a temporal instability. Moreover, note that the wavenumber $k_I = \omega/c_I$. That is, $k_I = k_{I0} + i\epsilon/c_I$, where $k_{I0} = \omega_0/c_I$. Therefore, for t fixed, $\lim_{x\to\infty} f(\boldsymbol{x})e^{i(k_I x-\omega t)} = 0$ provided $f(\boldsymbol{x})$ remains bounded. That is, if we want the wavefield to go to zero as $x \to \infty$, we must use the exponential term $e^{ik_I x}$ in combination with $e^{-i\omega t}$, and reject $e^{-ik_I x}$ as a possible solution. In *Proposition 4.5* we shall find that this choice of ω and hence of k is needed to ensure a unique solution. Once we have completed our calculation we may invoke analytic continuation[4] and, by letting $\epsilon \to 0$, recover the result for the case of real ω. This device of *setting* $\omega = \omega_0 + i\epsilon$ *to determine which wavefields are outgoing as* $x \to \infty$ *and hence to select which particular solutions to a problem give outgoing waves* is called *the principle of limiting absorption.* Note that the sign conventions leading to the evaluation of the contour integral, (4.14), and all previous contour integrals, are consistent with this principle. In particular we have implicitly used this principle when evaluating the contour integral, (4.14).

[4] Here we are invoking the *law of permanence of functional equations* (Hille, 1973).

4.5 Integral Representation: A Source Problem

Proposition 4.3. *Consider the region $\mathcal{R}_x$, shown in Fig. 4.1, with the region $\mathcal{R}$ absent, but $\mathcal{S}$ present. $\mathcal{S}$ is a bounded subregion containing time harmonic, body forces per unit mass, f_i, that excite the wavefield u_i in $\mathcal{R}_x$; u_i is the solution to*

$$(\lambda + \mu)\partial_i\partial_k u_k + \mu\partial_j\partial_j u_i + \rho\omega^2 u_i = -\rho f_i(\boldsymbol{x}), \qquad f_i(\boldsymbol{x}) = 0, \ \boldsymbol{x} \notin \mathcal{S}. \tag{4.30}$$

The wavefield satisfies the principle of limiting absorption as $x \to \infty$. After $\partial\mathcal{R}_x$ is sent to infinity, the solution u_i can be represented as

$$u_k(\boldsymbol{x}) = \int_S f_i(\boldsymbol{x}')u^G_{ik}(\boldsymbol{x} - \boldsymbol{x}')\, dV(\boldsymbol{x}'), \tag{4.31}$$

where u^G_{ik} is given by (4.7) and also satisfies the principle of limiting absorption.

Proof. Starting with the bounded region $\mathcal{R}_x$, we select as reciprocating wavefield 1 the triple (f_i, u_i, τ_{ij}) and as 2 the triple $[\delta(\boldsymbol{x}-\boldsymbol{x}')\hat{a}_i, u^G_{ik}(\boldsymbol{x}-\boldsymbol{x}')\hat{a}_k, \tau^G_{ijk}(\boldsymbol{x}-\boldsymbol{x}')\hat{a}_k]$, where u^G_{ik} is given by (4.7). Then (4.2) becomes

$$\int_{\mathcal{R}_x} \left[\delta(\boldsymbol{x} - \boldsymbol{x}')u_i(\boldsymbol{x})\hat{a}_i - f_i(\boldsymbol{x})u^G_{ik}(\boldsymbol{x} - \boldsymbol{x}')\hat{a}_k\right] \rho\, dV(\boldsymbol{x}) = I_k\hat{a}_k, \tag{4.32}$$

where

$$I_k = \int_{\partial\mathcal{R}_x} \left[u^G_{ik}(\boldsymbol{x} - \boldsymbol{x}')\tau_{ji}(\boldsymbol{x}) - u_i(\boldsymbol{x})\tau^G_{jik}(\boldsymbol{x} - \boldsymbol{x}')\right]\hat{n}_j\, dS(\boldsymbol{x}). \tag{4.33}$$

We shall subsequently show that $\lim_{x\to\infty} I_k = 0$. Therefore (4.32) reduces to

$$u_i(\boldsymbol{x}')\hat{a}_i = \int_S f_i(\boldsymbol{x})u^G_{ik}(\boldsymbol{x} - \boldsymbol{x}')\hat{a}_k\, dV(\boldsymbol{x}). \tag{4.34}$$

By inspection, $u^G_{ik}(\boldsymbol{x} - \boldsymbol{x}') = u^G_{ik}(\boldsymbol{x}' - \boldsymbol{x})$. Hence we are lead to (4.31), after a relabeling of the independent and integration variables.

Returning to the question of how I_k behaves, we use the asymptotic results of (4.22)–(4.26) to show that, as $x \to \infty$,

$$\begin{aligned} I_k \sim{} & \frac{1}{c_L^2}\frac{e^{ik_L x}}{4\pi x}\int_{\partial\mathcal{R}_x} [\tau_{ji}\hat{n}_j - ik_L(\lambda + 2\mu)u_i]\hat{n}_i\hat{n}_k e^{-ik_L\hat{\boldsymbol{n}}\cdot\boldsymbol{x}'}\, dS(\boldsymbol{x}') \\ & + \frac{1}{c_T^2}\frac{e^{ik_T x}}{4\pi x}\int_{\partial\mathcal{R}_x} [\tau_{ji}\hat{n}_j - ik_T\mu u_i](\delta_{ik} - \hat{n}_i\hat{n}_k)e^{-ik_T\hat{\boldsymbol{n}}\cdot\boldsymbol{x}'}\, dS(\boldsymbol{x}'). \end{aligned} \tag{4.35}$$

Note that we have relabeled the variables from those used in (4.33). When the principle of limiting absorption is invoked, $\lim_{x\to\infty} I_k = 0$. □

4.5.1 Notes

1. When thinking through the argument just given, note that two issues have been settled by using the principle of limiting absorption. First, reexamining the argument leading to the Green's tensor, recall that we demanded that the waves be outgoing. In place of the phrase *outgoing*, we now ask that the wavefield satisfy the principle of limiting absorption and thus approach zero as $x \to \infty$. Therefore we must choose the solution that behaves as e^{ikx} and not e^{-ikx}. The principle thus serves to determine the solution uniquely. Second, through its use, integrals such as I_k can be shown to approach zero without having to make detailed asymptotic estimates of how they behave as $x \to \infty$.

2. As an alternative to using the principle of limiting absorption, we could have demanded that

$$\lim_{x\to\infty} \int_{\partial\mathcal{R}_x} |[\tau_{ij}\hat{n}_j - ik_L(\lambda+2\mu)\,u_i]\hat{n}_i|^2\, dS(\boldsymbol{x}') = 0,$$

$$\lim_{x\to\infty} \int_{\partial\mathcal{R}_x} |[\tau_{ij}\hat{n}_j - ik_T\mu u_i](\delta_{ik} - \hat{n}_i\hat{n}_k)|^2 dS(\boldsymbol{x}') = 0. \tag{4.36}$$

This is one way of stating the elastodynamic radiation conditions. Note that they would only be satisfied by an outgoing wave. That is one having the form $f(\boldsymbol{x})e^{ik_Ix}$. From this it is possible to show that $\lim_{x\to\infty} I_k = 0$ (Achenbach et al., 1982).

3. Note that (4.31) asymptotically approaches the form

$$u_i \sim A_i(\boldsymbol{x})\frac{e^{ik_Lx}}{k_Lx} + B_i(\boldsymbol{x})\frac{e^{ik_Tx}}{k_Tx}, \qquad x \to \infty, \tag{4.37}$$

provided $\mathcal{S}$ is bounded. In this case $\mathcal{S}$ is said to be a compact source because the support for $f_i(\boldsymbol{x})$ is compact or bounded. Irrespective of the geometry of the source, provided it is compact, the wavefield must evolve into two separate spherical waves with radiation patterns given by A_i and B_i.

4.6 Integral Representation: A Scattering Problem

We continue to consider the region $\mathcal{R}_x$ shown in Fig. 4.1. Within it we imagine that the total wavefield $u_i^t = u_i^i + u_i$, where u_i^i is the wavefield excited by the sources in $\mathcal{S}$ in the absence of the subregion $\mathcal{R}$ and u_i is that scattered by $\mathcal{R}$. We next perform a slight of hand whereby we imagine that the subregion $\mathcal{S}$

is removed from $\mathcal{R}_x$ so that the scattered wavefield u_i does not interact with the sources in $\mathcal{S}$. Lastly, we consider $\mathcal{R}$ to be an empty cavity and ask that the traction vanish on $\partial\mathcal{R}$. We now formulate a boundary-value problem for u_i, with $t_i = -t_i^i$ on $\partial\mathcal{R}$ acting as a source. It is useful at this point to define $\mathcal{R}$ as an open region so that it does not contain its boundary points $\partial\mathcal{R}$. Then we define the region $\mathcal{R}_x\backslash\mathcal{R}$ as that contained in $\mathcal{R}_x$ excluding $\mathcal{R}$, but containing the boundary points $\partial\mathcal{R}$.

Proposition 4.4. *Consider the region $\mathcal{R}_x$ shown in Fig. 4.1 with the region $\mathcal{S}$ absent, but $\mathcal{R}$ present. Within $\mathcal{R}_x\backslash\mathcal{R}$ the time-harmonic wavefield u_i is a solution to (4.30), with the right-hand side set to zero. On $\partial\mathcal{R}$, u_i is unknown, but $\tau_{ji}\hat{n}_j = -\tau_{ji}^i\hat{n}_j$, where τ_{ji}^i is known. Moreover, u_i satisfies the principle of limiting absorption as $x \to \infty$. It follows then that as the radius $x \to \infty$, u_i can be represented as*

$$u_k(\boldsymbol{x}) = -\frac{1}{\rho}\int_{\partial\mathcal{R}} \left[u_{ik}^G(\boldsymbol{x}-\boldsymbol{x}')\tau_{ji}^i(\boldsymbol{x}') + u_i(\boldsymbol{x}')\tau_{jik}^{G\prime}(\boldsymbol{x}-\boldsymbol{x}')\right]\hat{n}_j\, dS(\boldsymbol{x}'). \quad (4.38)$$

The prime given τ_{jik}^G indicates that the $\boldsymbol{x}'$ is differentiated when $\tau_{jik}^{G\prime}$ is calculated by using (4.15).

Proof. Select as reciprocating wavefield 1 the triple $(0, u_i, \tau_{ij})$ and as 2 the triple $[\hat{a}_i\delta(\boldsymbol{x}-\boldsymbol{x}'), u_{ik}^G(\boldsymbol{x}-\boldsymbol{x}')\hat{a}_k, \tau_{ijk}^G(\boldsymbol{x}-\boldsymbol{x}')\hat{a}_k]$. Using (4.2) and proceeding as we did with *Proposition 4.3*, we arrive at

$$\rho u_i(\boldsymbol{x}')\hat{a}_i = \int_{\partial\mathcal{R}} (u_{ik}^G\tau_{ji} - u_i\tau_{jik}^G)\hat{a}_k\hat{n}_j\, dS(\boldsymbol{x}) + I_k\hat{a}_k. \quad (4.39)$$

Using the boundary condition on $\partial\mathcal{R}$, relabeling the independent and integration variables and noting that again $\lim_{x\to\infty} I_k = 0$, by the principle of limiting absorption, we obtain (4.38). □

4.6.1 Notes

1. Note, in (4.38), that $\tau_{jik}^{G\prime}$ is calculated from $u_{ik}^G(\boldsymbol{x}-\boldsymbol{x}')$ by differentiating with respect to $\boldsymbol{x}'$, the second argument of the combination $(\boldsymbol{x}-\boldsymbol{x}')$.

2. Note that (4.38) is *not a solution* to the boundary-value problem because the integral contains the unknown u_i evaluated on the boundary. In fact we have used the incorrect Green's tensor for this problem. We should have calculated a Green's tensor for which $\tau_{jik}^G\hat{n}_j = 0$ on $\partial\mathcal{R}$. This is usually far from easy to do. Nevertheless, (4.38) forms the starting point for many useful approximations.

3. The integral representation for $u_i(\boldsymbol{x})$ can be used to construct an integral equation, as *Problem 4.3* suggests. In forming the integral equation care must be taken because, as $\boldsymbol{x}$ approaches the surface, the Green's tensors become singular. Kevorkian (1993) gives an introduction to how to examine the consequences of these singularities within his discussion of integral equations in potential theory. Integral equations are sometimes easier to solve numerically than are differential ones, and their solution lends insight that direct numerical methods often do not. One book describing the numerical solution of integral equations is that by Delves and Mohamed (1985).

Problem 4.3 Integral Equation Problems

These are two dimensional, antiplane shear problems designed to ask the reader to derive directly, in this simpler context, several of the formulas given previously. In both the problems that follow, the elastic half-space is described by $\{(x_1, x_2)| -\infty < x_1 < \infty, -\infty < x_2 \leq 0\}$.

Problem 1. An elastic half-space is clamped at the surface $x_2 = 0$ by a rigid strip over $|x_1| < a$. This forces the particle displacement at the surface to go to zero there. For $|x_1| > a$ the surface is free of traction. An antiplane shear wave

$$u_3^i = Ae^{ikx_2} \tag{4.40}$$

is normally incident to the rigid strip and adjacent free surface. Determine an integral equation for the scattered wavefield.

Problem 2. Now assume that the region $|x_1| > a$ is clamped by rigid sheets, while within a slot $|x_1| < a$ the surface is free of traction. For the same incident wave, determine an integral equation for the scattered wavefield.

Outline of the Solution to Problem 1

1. Divide the total particle displacement u_3^t into the sum of two parts, namely, $u_3^t = u_3^i + u_3^s$, where u_3^s is the scattered wavefield. Then divide this latter wavefield again into two parts, namely, $u_3^s = u_3^r + u_3$, where u_3^r is the plane wave reflected from the surface with no strip present and u_3 is that caused by the presence of the rigid strip. Show that u_3 satisfies the time-harmonic wave equation and the following boundary conditions along $x_2 = 0$: $u_3 = -2A$ on $|x_1| < a$, and $\partial_2 u_3 = 0$ on $|x_1| > a$. How does u_3 behave at ∞? At $\pm a$?
2. Use the method of images to find the antiplane shear, Green's function $u_3^G(x_1, x_2|x_1', x_2')$, for the half-space satisfying a homogeneous boundary

condition $\partial_2 u_3^G = 0$ at $x_2 = 0$, and the principle of limiting absorption as $x \to \infty$. Note that what is needed is the Green's function for this problem. In contrast, the Green's tensor derived in *Proposition 4.2* and used in *Propositions 4.3* and *4.4* was that for a region without boundaries and not the correct Green's tensor for the scattering problem just discussed.

Hint. The reader may find the results of *Problem 4.1* useful.

3. Derive the reciprocity identity for two reciprocating, antiplane shear wavefields. Taking as one reciprocating wavefield the Green's function and as the other u_3, show that

$$u_3(x_1, x_2) = -\int_{-a}^{a} u_3^G(x_1, x_2|x_1', 0)I(x_1')dx_1', \tag{4.41}$$

where $I(x_1') = \partial_2' u_3(x_1', 0)$ and the prime indicates that the derivative is taken with respect to x_2'.

4. Use the boundary conditions satisfied by u_3 to derive the integral equation

$$2A = \frac{i}{2}\int_{-a}^{a} H_0^{(1)}(k|x_1 - x_1'|)I(x_1')dx_1'. \tag{4.42}$$

Note that $I(x_1)$ is the unknown to be solved for. After $I(x_1)$ is found, the representation (4.41) is used to calculate u_3 throughout the region of interest. Describe the nature of the singularities, in $[-a, a]$, possessed by the integrand.

Repeat the above steps for the second problem, making appropriate changes where necessary. In particular use a Green's function that satisfies $u_3^G = 0$ at $x_2 = 0$. Note that on this occasion the integrand is more singular than in the previous case. While one can work with singular integral equations, this particular equation can be reduced to one that is no more singular than that of the first problem (Sommerfeld, 1964a).

4.7 Uniqueness in an Unbounded Region

4.7.1 No Edges

Proposition 4.5. *Consider the region $\mathcal{R}_x$ shown in Fig. 4.1. Consider two time-harmonic wavefields (u_i^1, τ_{ij}^1) and (u_i^2, τ_{ij}^2) that satisfy*

$$\partial_j \tau_{ji}^{1,2} + \rho f_i + \rho\omega^2 u_i^{1,2} = 0. \tag{4.43}$$

in $\mathcal{R}_x \backslash \mathcal{R}$. The force $f_i \neq 0$ for $\mathbf{x} \in S$ but is zero elsewhere; ρ, λ, and μ are real parameters. The tractions on $\partial\mathcal{R}$ are equal; that is, $\tau_{ji}^1 \hat{n}_j = \tau_{ji}^2 \hat{n}_j$. The *wavefields are sufficiently continuous on ∂R that Gauss' theorem may be used. In accord with the principle of limiting absorption, $\omega = \omega_0 + i\epsilon$ so that both wavefields go to zero as $x \to \infty$. Then the two wavefields are identically equal in $\mathcal{R}_x \backslash \mathcal{R}$.*

Proof. Set $\mathbf{u} = \mathbf{u}^1 - \mathbf{u}^2$ and $\tau_{ij} = \tau^1_{ij} - \tau^2_{ij}$. Then (u_i, τ_{ij}) satisfies

$$\partial_k \tau_{kl} + \rho\omega^2 u_l = 0 \tag{4.44}$$

in $\mathcal{R}_x \backslash \mathcal{R}$. Moreover, (u^*_i, τ^*_{ij}) satisfies the same equation with ω^2 replaced by $(\omega^2)^*$. As was done previously, the superscript asterisk indicates the complex conjugate. Take the scalar product of (4.44) with u^*_i and the scalar product of the complex conjugate of (4.44) with u_i. Subtracting one from the other and using (1.3), the stress–strain relation, gives

$$[(\omega^2)^* - \omega^2]\rho u^*_l u_l = \partial_j(\tau_{ji} u^*_i - \tau^*_{ji} u_i). \tag{4.45}$$

Integrating this result gives

$$[(\omega^2)^* - \omega^2] \int_{\mathcal{R}_x \backslash \mathcal{R}} \rho u^*_l u_l dV = \int_{\partial\mathcal{R} \cup \partial\mathcal{R}_x} (\tau_{ji} u^*_i - \tau^*_{ji} u_i)\hat{n}_j dS. \tag{4.46}$$

The right-hand integral is zero because $\tau_{ij}\hat{n}_j = 0$ on $\partial\mathcal{R}$ and the integral over $\partial\mathcal{R}_x$ goes to zero as $x \to \infty$, by the principle of limiting absorption. Therefore the right-hand side equals zero. In addition, $\omega^* \neq \omega$, by the principle of limiting absorption, so that the left-hand integral must equal zero. The integrand is either everywhere zero or positive. If positive somewhere, then the integral cannot be zero, giving a contradiction. Therefore u_i and hence τ_{ij} must be zero throughout $\mathcal{R}_x \backslash \mathcal{R}$[5]. □

4.7.2 Edge Conditions

To prove uniqueness (or, in fact, to prove *Proposition 4.1* and hence all the subsequent propositions), it was essential that we be able to use Gauss' theorem. Thus the wavefields must have a certain measure of continuity on the surfaces. In the simpler proofs one asks that the partial derivatives be continuous apart from finite jumps (Courant and John, 1989; Kellogg, 1970). In Fig. 4.2 a sketch of the region $\mathcal{R}$, in cross section, with a lancet-shaped singularity whose edge is perpendicular to the page, is given. The wavefield is singular there. The

[5] Note that the analytic continuation of a function is unique. Hence, when the unique solution is determined by setting $\omega = \omega_0 + i\epsilon$, that uniqueness is not lost by taking the limit $\epsilon \to 0$.

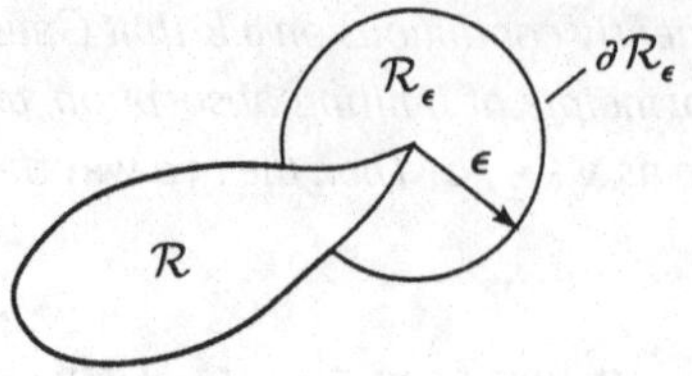

Fig. 4.2. The subregion $\mathcal{R}$ is drawn, in cross section, with a lancet-shaped singularity whose edge is perpendicular to the page. At such an edge the wavefield is singular. The edge is surrounded by a cylindrical region $\mathcal{R}_\epsilon$ with a radius ϵ. Both regions are contained within $\mathcal{R}_x$.

particle displacement is usually bounded even continuous, but its derivatives are singular. Thus Gauss' theorem is inapplicable near this edge without a certain amount of care being taken.

To investigate what happens at this edge, we surround it with a region $\mathcal{R}_\epsilon$ whose surface is $\partial\mathcal{R}_\epsilon$. This region does not intrude into (is disjoint from) $\mathcal{R}$. Again it is useful to define $\mathcal{R}$ and $\mathcal{R}_\epsilon$ as open regions that do not contain their boundary points. Gauss' theorem may be applied to the region $\mathcal{Q} = \mathcal{R}_x \backslash (\mathcal{R} \cup \mathcal{R}_\epsilon)$. The reader should look back at Fig. 4.1 and thereby recall that $\mathcal{R}$ and $\mathcal{R}_\epsilon$ are both contained within $\mathcal{R}_x$. The surface $\partial\mathcal{Q} = \partial\mathcal{R}_x \cup \partial(\mathcal{R} \cup \mathcal{R}_\epsilon)$. Now (4.46) is written as

$$[(\omega^2)^* - \omega^2] \int_{\mathcal{Q}} \rho u_l^* u_l dV = \int_{\partial\mathcal{Q}} (\tau_{ji} u_i^* - \tau_{ji}^* u_i) \hat{n}_j dS. \tag{4.47}$$

To complete the uniqueness proof we must ask additionally that

$$\lim_{\epsilon \to 0} \left[\Im \int_{\partial\mathcal{R}_\epsilon} \tau_{ji} u_i^* \hat{n}_j dS \right] = 0. \tag{4.48}$$

Then the right-hand side of (4.47) goes to zero subject to (4.48) being satisfied. And the uniqueness argument proceeds as before. A condition such as this is called an *edge condition.* However, it is more usual to enforce the edge condition by asserting the nature of the singularity that will be permitted at the edge so that (4.48) is satisfied. *It is important to realize that this singularity must be specified before the scattering problem is solved* and is as important in achieving a unique solution as is specifying a boundary condition or invoking the principle of limiting absorption.

Next we shall work through a specific case. However, for an edge, an element of surface area ΔS_ϵ of $\partial\mathcal{R}_\epsilon$ is $O(\epsilon)$. Moreover, typically $u_i \sim O(\epsilon^\alpha)$ and $\tau_{ij} \sim O(\epsilon^{\alpha-1})$, where $\alpha < 1$. Therefore, to satisfy (4.48), we demand $\alpha > 0$.

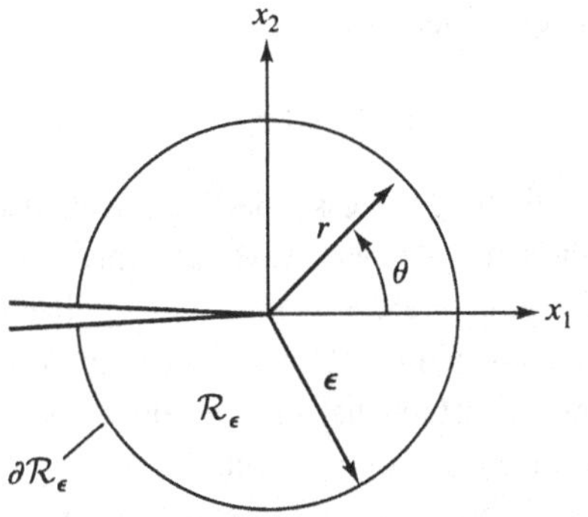

Fig. 4.3. The subregion $\mathcal{R}$ has been collapsed to a infinitesimally thin, semi-infinite slit or crack and $\partial\mathcal{R}$ becomes its two surfaces, upper and lower. The edge is perpendicular to the page, extending inward and outward to infinity. It is surrounded by a cylindrical region $\mathcal{R}_\epsilon$ with a radius ϵ that is smaller than a wavelength.

4.7.3 An Inner Expansion

We now consider a specific case of a scatterer with an edge. Figure 4.3 shows the region $\mathcal{R}$ collapsed to a infinitesimally thin, semi-infinite slit or crack defined by $\{(x_1, x_3)| - \infty < x_1 \leq 0, -\infty < x_3 < \infty\}$. Its surface $\partial\mathcal{R}$, the upper and lower surfaces of the crack, is free of traction. Though $\mathcal{R}$ is no longer bounded, our previous arguments are readily extended to this case. We wish to find the wavefield in the neighborhood of the edge to determine the nature of the wavefield's singularity there.

For simplicity, we take the wavefield to be an antiplane shear one. We seek a solution to (1.15), in the time-harmonic approximation and expressed in polar coordinates, namely

$$(1/r)\partial_r(r\partial_r u_3) + (1/r^2)\partial_\theta^2 u_3 + k^2 u_3 = 0, \tag{4.49}$$

where k is the wavenumber. At $\theta = \pm\pi$, $r \neq 0$, $\partial_\theta u_3 = 0$. We do not intend to solve a global problem but merely to explore possible solutions in the neighborhood of $r = 0$. Therefore we should use a coordinate expansion in kr for u_3 beginning with $(kr)^\alpha$, $0 < \alpha < 1$, with the assumption that $kr \to 0$. However, a bit more insight is gained by introducing the length scale ϵ shown in Fig. 4.3, where $k\epsilon \ll 1$. Setting $\rho = r/\epsilon$ and $w = u_3/U$ (U is a maximum particle displacement), we reexpress (4.49) in terms of (ρ, θ) and w. Asymptotically expanding w as

$$w(\rho, \theta) \sim (k\epsilon)^\alpha \sum_{n\geq 0} w_n(\rho, \theta)(k\epsilon)^n, \tag{4.50}$$

we find that the lowest-order term satisfies

$$(1/\rho)\partial_\rho(\rho\partial_\rho w_0) + (1/\rho^2)\partial_\theta^2 w_0 = 0. \tag{4.51}$$

The wavefield near the edge is essentially equivalent to the static field. This further illustrates the principle that a wave has to propagate several wavelengths, freeing itself of its source, before its propagating character becomes manifest. An expansion such as (4.50) is called an inner expansion. This expansion will contain an unknown constant that can be found by matching it to an outer expansion. The inner expansion connects the source – in this case the crack tip – with the propagating wavefield, the outer expansion, a few wavelengths away. This method of analysis is called *matching asymptotic expansions* (Hinch, 1991; Holmes, 1995). We shall continue with this analysis in Section 5.6.

Subject to the boundary condition stated previously, (4.51), has solutions of the form

$$w_0(\rho, \theta) = (A\rho^\beta + B\rho^{-\beta}) \sin \beta\theta, \tag{4.52}$$

where $\beta = (2n+1)/2$, $n = 0, 1, 2, \ldots$ and A and B are undetermined constants. To satisfy (4.48), $B = 0$ and $\alpha = \beta = \frac{1}{2}$. Larger values of β also give acceptable solutions, but this is the minimum such value and hence determines the most singular term allowable. Thus $ku_3 \sim (kr)^{1/2} A \sin(\theta/2) + \mathrm{O}[(kr)^{3/2}]$ as $kr \to 0$.

Therefore, for a problem with the geometry of Fig. 4.3, we must specify, when we first formulate the problem, that $ku_3 = \mathrm{O}[(kr)^{1/2}]$ as $kr \to 0$ to ensure that we arrive at a unique solution.

4.8 Scattering From an Elastic Inclusion in a Fluid

To convince the reader of just how powerful and general the foregoing results can be, in this closing section we extend slightly these results to develop an expression for the total wavefield present in an ideal fluid that contains an elastic inclusion. We again take the time dependence as harmonic. The elastic inclusion, in contrast to the empty cavity of *Proposition 4.4*, is penetrable. Waves enter it, reverberate within it, and then reradiate. This calculation is based on the work of Wickham (1992) and Leppington (1995).

The elastic inclusion occupies a region $\mathcal{R}$ with surface $\partial\mathcal{R}$. The unit normal $\hat{\boldsymbol{n}}$ points out of $\mathcal{R}$ (in contrast with our previous convention). We indicate by $\partial\mathcal{R}^+$ the surface approached from outside $\mathcal{R}$ and by $\partial\mathcal{R}^-$ that approached from within. Figure 4.1 indicates the geometry, if we imagine that the radius $x \to \infty$ and the region $\mathcal{S}$ is absent. To indicate when a position vector $\boldsymbol{x}$ identifies a point in $\mathcal{R}$, we define the function $\chi(\boldsymbol{x})$ as

$$\chi(\boldsymbol{x}) = \begin{cases} 1, & \boldsymbol{x} \in \mathcal{R}, \\ 0, & \text{otherwise} \end{cases} \tag{4.53}$$

We treat the inclusion as a linearly elastic solid embedded in a linearly elastic (ideal) fluid. The equations of motion for an elastic fluid are given by those of linear elasticity when the shear coefficient is set to zero. The elastic fluid is compressible but cannot withstand shearing. The equation of motion for the particle displacement $\boldsymbol{u}$ is

$$\bar{\lambda}\partial_i\partial_k u_k + \omega^2\bar{\rho}u_i = -\bar{\rho}\, f_i, \qquad \boldsymbol{x} \notin \mathcal{R}. \tag{4.54}$$

The stress tensor is given by $\tau_{ij} = -p\delta_{ij}$, where p, the acoustic pressure, equals $-\bar{\lambda}\partial_k u_k$. As before, $\boldsymbol{f}$ is the body force per unit mass. The parameters $\bar{\rho}$ and $\bar{\lambda}$ are the density and bulk modulus, respectively, of the elastic fluid. Lastly, there is another equation of motion, namely $\nabla \wedge \boldsymbol{u} = 0$, arising because the motion is irrotational.

We write the equation of motion for the particle displacement $\boldsymbol{u}$ in the elastic solid in such a way that the solid's inertial and elastic properties appear as a body force within the region $\mathcal{R}$. The equation is thus written as

$$\bar{\lambda}\partial_i\partial_k u_k + \omega^2\bar{\rho}u_i = -\partial_j\sigma_{ji} - \omega p_i, \qquad \boldsymbol{x} \in \mathcal{R}. \tag{4.55}$$

The right-hand terms are given by

$$\sigma_{ij} = (\lambda - \bar{\lambda})\delta_{ij}\partial_k u_k + \mu(\partial_i u_j + \partial_j u_i), \tag{4.56}$$

$$p_i = \omega(\rho - \bar{\rho})u_i. \tag{4.57}$$

Next we imagine that an incident wavefield $\boldsymbol{u}^i$ scatters from the elastic inclusion, giving rise to the scattered wavefield $\boldsymbol{u}^s$. The total wavefield $\boldsymbol{u}(\boldsymbol{x}) = [1 - \chi(\boldsymbol{x})]\boldsymbol{u}^i(\boldsymbol{x}) + \boldsymbol{u}^s(\boldsymbol{x})$. Across the surface $\partial\mathcal{R}$ the normal components of particle displacement and traction are continuous and the tangential components of traction vanish. We express these conditions as

$$[u_j^s]\hat{n}_j = -u_j^i\hat{n}_j, \quad [\partial_k u_k^s]\delta_{ij}\hat{n}_j = -\partial_k u_k^i\delta_{ij}\hat{n}_j + \sigma_{ji}\hat{n}_j/\bar{\lambda}. \tag{4.58}$$

The symbol $[\cdots]$ indicates the jump in going from $\partial\mathcal{R}^+$ to $\partial\mathcal{R}^-$.

Taken together, (4.55)–(4.58) capture the concept that the elastic solid is an inclusion within a host material, the elastic fluid, with the departure of the properties of the solid from those of the fluid expressed as body force terms on the right-hand side of (4.55) and in the conditions across $\partial\mathcal{R}$, (4.58).

Our goal is to use *Propositions 4.2–4.4*, or a slight variation of them, to give an integral representation for $\boldsymbol{u}$ as a volume integral over $\mathcal{R}$ that contains σ_{ij} and p_i. To do so we need the Green's tensor for the elastic fluid. This satisfies an equation analogous to (4.6), namely.

$$\bar{\lambda}\partial_i\partial_k u^G_{km} + \bar{\rho}\omega^2 u^G_{im} = -\bar{\rho}\delta_{im}\delta(\boldsymbol{x}). \tag{4.59}$$

It is calculated exactly as outlined in *Proposition 4.2* and is given by

$$u^G_{ik} = 1/k^2c^2[-\partial_i\partial_k(e^{ikx}/4\pi x) + \partial_m\partial_m(\delta_{ik}/4\pi x)]. \tag{4.60}$$

Note that this is not the usual Green's function for linear acoustics, which function is the response to a mass flux at a single point. However, it is the correct Green's tensor when one considers the fluid as elastic and seeks its response to a force acting at a single point. For a point force in the direction $\hat{\mathbf{a}}$, the reader is asked to verify that $u^G_{ik}\hat{a}_k$ is irrotational, provided $\boldsymbol{x} \neq 0$.

Next we apply arguments identical to those used in *Propositions 4.4* and then in *4.3*. The reciprocity relation is applied to the complement of $\mathcal{R}$. The Green's tensor is one reciprocating wavefield and $\boldsymbol{u}^s$ the second. The outcome is

$$u^s_m(\boldsymbol{x})[1-\chi(\boldsymbol{x})] = -c^2\int_{\partial\mathcal{R}+}[u^G_{jm}(\boldsymbol{x}-\boldsymbol{x}')\partial_k u^s_k(\boldsymbol{x}') - u^s_j(\boldsymbol{x}')\partial'_k u^G_{km}(\boldsymbol{x}-\boldsymbol{x}')]\hat{n}_j dS(\boldsymbol{x}'). \tag{4.61}$$

The wavespeed $c = (\bar{\lambda}/\bar{\rho})^{1/2}$. The prime indicates that the derivative is taken with respect to $\boldsymbol{x}'$. The unit normal $\hat{\boldsymbol{n}}$ points out of $\mathcal{R}$ rather than out of its compliment.

Next consider the region $\mathcal{R}$. Arguments that combine those used in both *Propositions 4.3* and *4.4* are used. Again the reciprocity relation is applied, but now within $\mathcal{R}$ itself. The Green's tensor is one reciprocating wavefield and $\boldsymbol{u}^s$ the second. The outcome is

$$\begin{aligned} u^s_m(\boldsymbol{x})\chi(\boldsymbol{x}) = {} & \frac{1}{\bar{\rho}}\int_{\mathcal{R}}[\partial_i\sigma_{ij}(\boldsymbol{x}') + \omega p_j(\boldsymbol{x}')]u^G_{jm}(\boldsymbol{x}-\boldsymbol{x}')dV(\boldsymbol{x}') \\ & + c^2\int_{\partial\mathcal{R}-}[u^G_{jm}(\boldsymbol{x}-\boldsymbol{x}')\partial_k u^s_k(\boldsymbol{x}') \\ & - u^s_j(\boldsymbol{x}')\partial'_k u^G_{km}(\boldsymbol{x}-\boldsymbol{x}')]\hat{n}_j dS(\boldsymbol{x}'). \end{aligned} \tag{4.62}$$

Recall that within $\mathcal{R}$, $\boldsymbol{u} = \boldsymbol{u}^s$. We add (4.61) and (4.62) to obtain the following representation for $\boldsymbol{u}^s$.

$$u^s_m(\boldsymbol{x}) = \frac{1}{\rho}\int_{\mathcal{R}} [\partial_i\sigma_{ij}(\boldsymbol{x}') + \omega p_j(\boldsymbol{x}')]u^G_{jm}(\boldsymbol{x}-\boldsymbol{x}')dV(\boldsymbol{x}') - c^2\int_{\partial\mathcal{R}} \left[u^G_{jm}(\boldsymbol{x}-\boldsymbol{x}')\left[\partial_k u^s_k\right](\boldsymbol{x}') - \left[u^s_j\right](\boldsymbol{x}')\partial'_k u^G_{km}(\boldsymbol{x}-\boldsymbol{x}')\right]\hat{n}_j dS(\boldsymbol{x}'). \tag{4.63}$$

As it stands, this is a remarkable expression. First note that the departure of the properties of the elastic solid from the host elastic fluid appears as a body force term in the integral over $\mathcal{R}$. Second note that the jumps $[u^s_j]\hat{n}_j$ and $[\partial_k u^s_k]\delta_{ij}\hat{n}_j$, given by (4.58), are present in the integral over $\partial\mathcal{R}$. The superscripts $\pm$ on $\partial\mathcal{R}$ have served their purpose and are omitted from now on.

Now we imagine that the region $\mathcal{R}$ contains only the incident wavefield $\boldsymbol{u}^i$ and use the reciprocity relation along with the Green's tensor to write

$$u^i_m(\boldsymbol{x})\chi(\boldsymbol{x}) = c^2\int_{\partial\mathcal{R}} \left[u^G_{im}(\boldsymbol{x}-\boldsymbol{x}')\partial_k u^i_k(\boldsymbol{x}')\delta_{ij} - u^i_j(\boldsymbol{x}')\partial'_k u^G_{km}(\boldsymbol{x}-\boldsymbol{x}')\right]\hat{n}_j dS(\boldsymbol{x}'). \tag{4.64}$$

This in itself is a interesting result. It is called an *extinction theorem*,[6] and it indicates that the surface integral vanishes for observation points $\boldsymbol{x}$ outside $\mathcal{R}$.

We now subtract (4.63) from (4.64) to obtain

$$u_m(\boldsymbol{x}) = u^i_m(\boldsymbol{x}) + \frac{1}{\rho}\int_{\mathcal{R}} [\partial_i\sigma_{ij}(\boldsymbol{x}') + \omega p_j(\boldsymbol{x}')]u^G_{jm}(\boldsymbol{x}-\boldsymbol{x}')dV(\boldsymbol{x}') - \frac{1}{\rho}\int_{\partial\mathcal{R}} \left[u^G_{im}(\boldsymbol{x}-\boldsymbol{x}')\sigma_{ij}(\boldsymbol{x}')\right]n_j dS(\boldsymbol{x}'). \tag{4.65}$$

We are almost done, but we are still left with a surface integral. Noting the divergence term in the volume integral, we make one more application of Gauss' theorem to give

$$u_m(\boldsymbol{x}) = u^i_m(\boldsymbol{x}) - \frac{1}{\rho}\int_{\mathcal{R}} \left[\sigma_{ij}(\boldsymbol{x}')\partial_i u^G_{jm}(\boldsymbol{x}-\boldsymbol{x}') - \omega p_j(\boldsymbol{x}')u^G_{jm}(\boldsymbol{x}-\boldsymbol{x}')\right]dV(\boldsymbol{x}'). \tag{4.66}$$

This is a very satisfying outcome. We have, through successive applications of the Green's tensor in combination with the reciprocity relation, shown that the

[6] The expressions (4.61) and (4.62) are also examples of extinction theorems.

total wavefield $\boldsymbol{u}$ outside $\mathcal{R}$ equals the incident wavefield $\boldsymbol{u}^i$ plus a scattered wavefield $\boldsymbol{u}^s$ whose source is the departure of the inertial and elastic properties of the solid inclusion from those of its host, the elastic fluid. However, (4.66) is not a solution to the boundary-value problem for $\boldsymbol{u}^s$. Just as with (4.38), the integral contains unknown terms. These are σ_{ij} and p_i, which cannot be calculated without knowing $\boldsymbol{u}$ within $\mathcal{R}$. Thus this expression is only a representation and not a solution to the problem. However, it can be made the basis for deriving a system of integral equations for σ_{ij} and p_i. One uses (4.66) in (4.56) and (4.57) to enforce self-consistency (Leppington, 1995).

References

Achenbach, J.D. 1973. *Wave Propagation in Elastic Solids*, pp. 96–110. Amsterdam: North-Holland.

Achenbach, J.D., Gautesen, A.K., and McMaken, H. 1982. *Ray Methods for Waves in Elastic Solids*, pp. 22–27 and 34–38. Boston: Pitman.

Courant, R. and John, F. 1989. *Introduction to Calculus and Analysis*, Vol. II, pp. 597–602. New York: Springer.

deHoop, A.T. 1995. *Handbook of Radiation and Scattering of Waves*. London: Academic.

Delves, L.M. and Mohamed, J.L. 1985. *Computational Methods for Integral Equations*. Cambridge: University Press.

Friedlander, F.G. 1958. *Sound Pulses*. Cambridge: University Press.

Hille, E. 1973. *Analytic Function Theory*, Vol. II, pp. 31–36. New York: Chelsea.

Hinch, E.J. 1991. *Perturbation Methods*, pp. 52–101. Cambridge: University Press.

Holmes, M.H. 1995. *Introduction to Perturbation Methods*, pp. 47–104. New York: Springer.

Hudson, J.A. 1980. *The Excitation and Propagation of Elastic Waves*, pp. 106–109. New York: Cambridge.

Kellogg, O.D. 1970. *Foundations of Potential Theory*, pp. 84–121. New York: Frederick Ungar.

Kevorkian, J. 1993. *Partial Differential Equations*, pp. 64–75 and 94–99. New York: Chapman and Hall.

Leppington, S.J. 1995. *The Scattering of Sound by a Fluid-Loaded, Semi-Infinite Thick Elastic Plate*, pp. 75–105. Manchester: Ph.D. Dissertation, The University of Manchester.

Sommerfeld, A. 1964a. *Optics, Lectures on Theoretical Physics*, Vol. IV, pp. 273–289. Translated by O. LaPorte and P.A. Moldauer. New York: Academic.

Sommerfeld, A. 1964b. *Partial Differential Equations in Physics, Lectures on Theoretical Physics*, Vol. VI, pp. 84–101. Translated by E.G. Straus. New York: Academic.

Wickham, G.R. 1992. A polarization theory for scattering of sound at imperfect interfaces. *J. Nondestr. Eval.* **11**: 199–210.

5

Radiation and Diffraction

Synopsis

Chapter 5 summarizes the basic propagation processes that are encountered when studying radiation or edge diffraction. Three problems of progressive difficulty are studied. We begin by calculating the transient, antiplane radiation excited by a line source at the surface of a half-space. The Cagniard–deHoop method is used to invert the integral transforms. We then return to considering how plane waves and a knowledge of their interactions can be used to construct more general wavefields. We calculate the time harmonic, inplane radiation, from a two-dimensional center of compression buried in a half-space. Plane-wave spectral techniques are used and the resulting integrals are approximated by the method of steepest descents. This method is discussed in detail. Lastly, we extend our knowledge of plane-wave interactions by calculating the diffraction of a time harmonic, plane, antiplane shear wave by a semi-infinite slit or crack. This problem is solved exactly by using the Wiener–Hopf method and approximately by using matched asymptotic expansions. An Appendix describing the reduction of the diffraction integral to Fresnel integrals is included.

5.1 Antiplane Radiation into a Half-Space

We consider an elastic half-space. The x_1 coordinate stretches along its surface and the positive x_2 coordinate extends into the interior. At the origin a line load is applied to an otherwise traction-free surface. The line load is a tangentially acting force very localized in x_1 and directed from $-\infty$ to ∞ in the x_3 direction. The equations of motion are given by (1.15) combined with (1.14). On the surface $x_2 = 0$,

$$\mu \partial_2 u_3 = -\mu \delta(x_1) f(t). \tag{5.1}$$

The half-space is quiescent for $t < 0$ so that $f(t) \equiv 0$ and $u_3 \equiv 0$ for $t < 0$. The waves are outgoing from the source. As we have done previously, the subscript T is dropped.

5.1.1 The Transforms

Cagniard (1962) developed a very instructive way to invert the integral transforms arising in transient wave problems. DeHoop (1960) popularized Cagniard's method by making it more readily understood. The essence of the technique, now referred to as the *Cagniard–deHoop technique*, is the mapping of the phase term, and subsequently the integration contour, of the spatial inverse transform into a form that allows one to immediately identify the inverse temporal transform. What is most satisfying about the technique is that it shifts the burden of inverting the transform to understanding a mapping that primarily affects the phase term, the amplitude term being adjusted almost as an afterthought. And the transform merely provides a shell within which the mapping lives.

Arguably, the simplest way to understand the technique is to use a one-sided Laplace transform over time and a two-sided one over space. First, the one-sided transform over time (1.31) is taken, with the restriction that the transform variable p remain real and positive. This is not a serious restriction because we shall not need to invert this transform. Moreover, if necessary, we can analytically continue the transform into the complex p plane. Second, we take the two-sided Laplace transform over the spatial coordinate; that is,

$$^*\bar{u}_3(\alpha, x_2, p) = \int_{-\infty}^{\infty} e^{-p\alpha x_1}\bar{u}_3(x_1, x_2, p)\,dx_1, \tag{5.2}$$

where the overbar indicates the temporal transform. Note that p is used to scale the spatial transform variable α; α is complex. Though we have not used this transform previously, it is only a slight modification of the Fourier transform of (1.43). The inverse of (5.2) is

$$\bar{u}_3(x_1, x_2, p) = \frac{p}{2\pi i}\int_{-\epsilon-i\infty}^{\epsilon+i\infty} e^{p\alpha x_1}\,{}^*\bar{u}_3(\alpha, x_2, p)\,d\alpha, \tag{5.3}$$

where $\epsilon \geq 0$.

Applying the transforms over t and x_1 leads to the ordinary differential equation

$$d^{2*}\bar{u}_3/dx_2^2 - p^2\chi^{2\,*}\bar{u}_3 = 0, \qquad \chi = (s^2 - \alpha^2)^{1/2}. \tag{5.4}$$

The slowness $s = 1/c$ is used instead of the wavespeed c. The solution must be selected in such a way that the outgoing condition is enforced. This can be achieved by asking that[1] $\Re(\chi) \geq 0\ \forall\alpha$ and taking as the solution $^*\bar{u}_3 = A(\alpha, p)e^{-p\chi x_2}$. Enforcing the transformed boundary condition at $x_2 = 0$ gives $A(\alpha, p) = \bar{f}(p)/(p\chi)$. Inverting in α by using (5.3) gives

$$\bar{u}_3(x_1, x_2, p) = \frac{\bar{f}(p)}{2\pi i}\int_{-\epsilon-i\infty}^{\epsilon+i\infty} e^{p\alpha x_1}\frac{e^{-p\chi x_2}}{\chi}\,d\alpha. \tag{5.5}$$

Next, inverting in time gives

$$u_3(x_1, x_2, t) = f(t) * I(x_1, x_2, t), \tag{5.6}$$

where the centered asterisk indicates a convolution in t. Hence, we are left with inverting the integral

$$\bar{I}(x_1, x_2, p) = \frac{1}{2\pi i}\int_{-\epsilon-i\infty}^{\epsilon+i\infty} e^{p\alpha x_1}\frac{e^{-p\chi x_2}}{\chi}\,d\alpha. \tag{5.7}$$

The ϵ has been added to the inversion contour in (5.3), and hence (5.7), so that should the contour be rotated to give a standard Fourier transform, the contour will pass above and below the branch cuts in such a way that $-p\chi$, with $p = -i\omega$ and $\omega > 0$, will give an outgoing wave for $x_2 > 0$ in the integrand of (5.7).

5.1.2 Inversion

To make further progress we must decide how to cut the α plane. The cuts are not the same as those indicated previously, because we are now using a Laplace rather than a Fourier transform. Nevertheless the reasoning is identical to that given in Section 3.4.4. Note that $\chi = -i\gamma$, where γ is given by (3.51). Thus, for $\Re(\chi) \geq 0\ \forall\ \alpha$, $\Im(\gamma) \geq 0\ \forall\alpha$ and the branch cuts shown in Fig. 5.1 follow. Here χ is given as

$$\chi = (r_1 r_2)^{1/2} e^{i\theta_1/2} e^{i\theta_2/2} e^{-i\pi/2}, \tag{5.8}$$

where (r_i, θ_i) are defined in Fig. 5.1. The angle $\theta_1 \in (0, 2\pi)$ and $\theta_2 \in (-\pi, \pi)$. Note that the sign of $\Im(\chi)$ varies from quadrant to quadrant, as indicated in the caption to Fig. 5.1.

[1] In the absence of propagation, we are asking that the components of the disturbance decay toward infinity. For p that is real and positive, this choice will give us the freedom to distort, almost anywhere, the spatial inversion contour in the complex α plane.

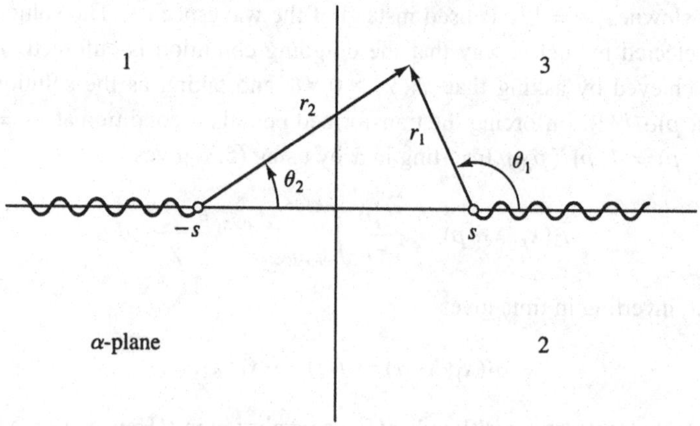

Fig. 5.1. The complex α plane cut so that $\Re(\chi) \geq 0 \, \forall \, \alpha$. In quadrants 1 and 2, $\Im(\chi) > 0$, and in quadrants 3 and 4, $\Im(\chi) < 0$. The magnitude r_i and argument θ_i for each radical are shown. Contrast this figure with Fig. 3.7.

We start by mapping the variable of integration from α to t, without assigning any physical meaning to t. That is,we set

$$t(\alpha) = -\alpha x_1 + \chi x_2. \tag{5.9}$$

The variable t traces out a contour in the complex t plane as α ranges from $(-\epsilon - i\infty)$ to $(\epsilon + i\infty)$. However, rather than look at the t plane, we find that it is simpler to continue examining the α plane. Setting $x_1 = r\cos\theta$ and $x_2 = r\sin\theta$, where $r = (x_1^2 + x_2^2)^{1/2}$, we use (5.9) to find α as a function of t. A quadratic equation for α is arrived at, and when solved gives the two roots

$$\alpha_\pm = -\frac{t}{r}\cos\theta \pm i\sin\theta\left(\frac{t^2}{r^2} - s^2\right)^{1/2}. \tag{5.10}$$

We next ask, along what contour in the α plane is t real and positive, and can we distort the current contour to that one? The answer to the latter question is yes, for $x_2 > 0$, because, with p real and positive, the integral (5.7) is convergent throughout the particular Riemann sheet we are working on. Figure 5.2 indicates the answer to the former question. The original contour is shown by the heavy, dashed line and the new one by the solid line. The new contour has two branches evinced by the subscripts plus and minus attached to the α. Eliminating the parameter t shows that the curve is a hyperbola with asymptotes $\Im(\alpha^\pm)/\Re(\alpha^\pm) = \mp\tan\theta$. The parameter t starts at sr, where $\alpha_\pm = -s\cos\theta$, and goes to ∞ along each branch $\alpha_\pm$. This contour is the Cagniard–deHoop

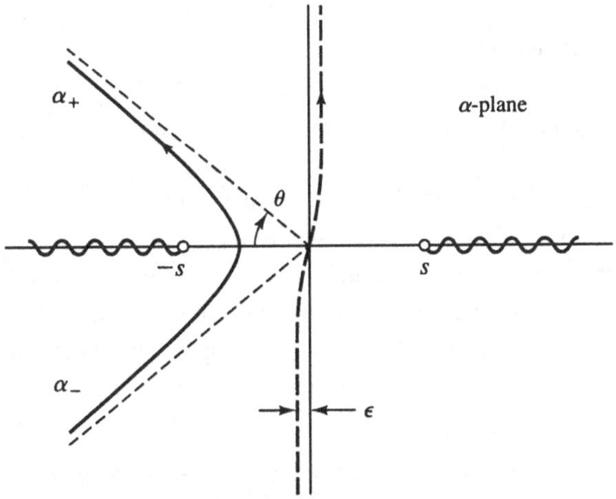

Fig. 5.2. A sketch of the original contour (heavy, dashed line) and the Cagniard–deHoop contour (solid line) in the complex α plane. Also shown are the asymptotes (light, dashed lines) to the Cagniard–deHoop contour and the angle θ. Note how the angle θ is defined.

contour. One last term is needed, namely

$$\frac{d\alpha_\pm}{dt} = -\frac{1}{r}\cos\theta \pm i\frac{t\sin\theta}{r^2[(t^2/r^2) - s^2]^{1/2}}. \tag{5.11}$$

Equation (5.7) can now be written as

$$\bar{I}(x_1, x_2, p) = \frac{1}{2\pi i}\int_{sr}^{\infty}\left[\frac{e^{-pt}}{(s^2 - \alpha_+^2)^{1/2}}\frac{d\alpha_+}{dt} - \frac{e^{-pt}}{(s^2 - \alpha_-^2)^{1/2}}\frac{d\alpha_-}{dt}\right]dt. \tag{5.12}$$

Noting that one component of the integrand is the complex conjugate of the other, we rewrite the integral as

$$\bar{I}(x_1, x_2, p) = \frac{1}{\pi}\int_{sr}^{\infty} e^{-pt}\Im\left[\frac{d\alpha_+/dt}{(s^2 - \alpha_+^2)^{1/2}}\right]dt. \tag{5.13}$$

Thus, by inspection,

$$I(x_1, x_2, t) = H(t - sr)\frac{1}{\pi}\Im\left[\frac{d\alpha_+/dt}{(s^2 - \alpha_+^2)^{1/2}}\right], \tag{5.14}$$

where $H(x)$ is the Heaviside function.

The burden of the inversion has rested with the mapping from the α plane to the t plane by using (5.9) and its inverse (5.10). Recall, however, that a convolution integral (5.6) must still be evaluated. Moreover, the example just given is a particularly simple instance of this mapping, and indeed Cagniard (1962) should be explored for more complicated cases.

It is also of interest to note the form of (5.13). It is precisely of the form needed to approximate it, for large p, using *Watson's lemma*, something we discuss in Section 5.3.1. With a bit of exploring, we should find that in doing so the leading term is determined by the nature of the singularity in the $d\alpha_+/dt$ of the integrand. Using a Tauberian theorem (van der Pol and Bremmer, 1950), we can construct an approximation to $I(x_1, x_2, t)$ for t near sr. This is called a *wavefront approximation.* This technique is described, within the context of elastic waves, by Knopoff and Gilbert (1959) and is used extensively by Harris (1980a,1980b). It is also briefly explored in *Problem 5.3.*

Problem 5.1 Half-Plane and Strip Problems

Problem 1. Consider the half-space $\{(x_1, x_2)| -\infty < x_1 < \infty, x_2 \geq 0\}$. It is subjected, at $x_2 = 0$, to the antiplane traction

$$\tau_{23} = -\mu\, H(x_1) df/dt \tag{5.15}$$

$f(t) \equiv 0$ for $t < 0$. Determine the particle displacement $u_3(x_1, x_2, t)$ in the interior by using the Cagniard–deHoop technique to invert the transforms. The integrand of the inverse spatial transform will have a pole and two branch points. Identify the waves represented by the pole term and by the integral taken along the Cagniard–deHoop contour.

Problem 2. A simple variation to *Problem 1* is to consider exactly the same problem except that (5.15) is replaced by

$$\tau_{23} = -\mu\, df/dt\, [H(x_1 + a) - H(x_1 - a)], \tag{5.16}$$

where a is a constant. Determine $u_3(x_1, x_2, t)$ for this case.

5.2 Buried Harmonic Line of Compression I

Finding the response of an elastic half-space to a point or line load applied either at its surface or buried just below it is possibly the most studied problem in elastic waves. It is usually called *Lamb's problem.* Ewing et al. (1957) and Cagniard (1962) are primary references to these problems, though all the books on elastic waves cited previously discuss them. We limit our discussion to calculating the response of an elastic half-space to a time-harmonic line of compression (a two-dimensional center of compression) by using the plane-wave spectral approach.

We consider an elastic half-space, identical to that discussed in Section 5.1, with a line of compression located at $(0, h)$. A source of this kind was introduced in Section 4.3.1 and explored further in *Problem 4.2.* The outcome of that problem suggests that the present calculations are most easily begun by working with potentials. The potentials and their governing equations are given by (1.19)–(1.22). The vector potential $\boldsymbol{\psi} = \psi \hat{\mathbf{e}}_3$, so that the divergence condition is automatically satisfied. The potentials are divided into two parts, namely $\varphi^t = \varphi^i + \varphi$ and $\psi^t = \psi$. The terms φ^t and ψ^t are the total compressional and shear potentials, respectively. The term φ^i is that excited by the line of compression in the absence of the traction-free surface and the terms φ and ψ are the compressional and shear potentials scattered from the surface. The source is time harmonic, and we suppress that dependence.

The term φ^i satisfies

$$\partial_\alpha \partial_\alpha \varphi^i + k_L^2 \varphi^i = F_0 \delta(x)\delta(y-h), \tag{5.17}$$

an equation almost identical to (4.17). The constant F_0 has the dimensions of length squared. It is sometimes useful to set $F_0 = A/k_L^2$, where A is a dimensionless constant. The potentials φ and ψ satisfy the same equation, but with no source term on the right-hand side, and in the equation governing ψ, with k_L replaced by k_T. Note that (4.18) gives the solution to (5.17). Written here using a different notation, the solution is

$$\varphi^i = \frac{-iF_0}{4\pi} \int_{C_\beta} e^{i(\beta x_1 + \gamma_L |x_2 - h|)} \frac{d\beta}{\gamma_L}. \tag{5.18}$$

The radical $\gamma_L = (k_L^2 - \beta^2)^{1/2}$. Note that $\gamma_L = i\gamma$, where γ is the radical defined by (3.51)[2]. We ask that $\Im(\gamma_L) \geq 0 \ \forall \beta$. The radical $\gamma_T = (k_T^2 - \beta^2)^{1/2}$, which will soon be needed, is defined in the same way. This leads to the complex plane structured as indicated in Fig. 5.3 (imagine the branch cuts for γ_L and γ_T lying on top of one another). The contour C_β is also shown.

The total particle displacement $\mathbf{u}^t = \mathbf{u}^i + \mathbf{u}$, the incident plus scattered wavefields. For $x_2 < h$, $\mathbf{u}^i$ is given by

$$\mathbf{u}^i = \frac{F_0}{4\pi} \int_{C_\beta} (\beta \hat{\mathbf{e}}_1 - \gamma_L \hat{\mathbf{e}}_2)\, e^{i[\beta x_1 + \gamma_L (h - x_2)]} \frac{d\beta}{\gamma_L}. \tag{5.19}$$

We next introduce the Sommerfeld transformation $\beta = k_L \cos\alpha,\ \gamma_L = k_L \sin\alpha$ first introduced in (2.37). Then (5.18) becomes

$$\mathbf{u}^i = -\frac{k_L F_0}{4\pi} \int_C \hat{\mathbf{p}}_0(\alpha) e^{ik_L \hat{\mathbf{p}}_0 \cdot \mathbf{x}} e^{ik_L h \sin\alpha}\, d\alpha, \tag{5.20}$$

[2] γ_L is the radical noted at the end of Section 3.44 and there labeled ξ.

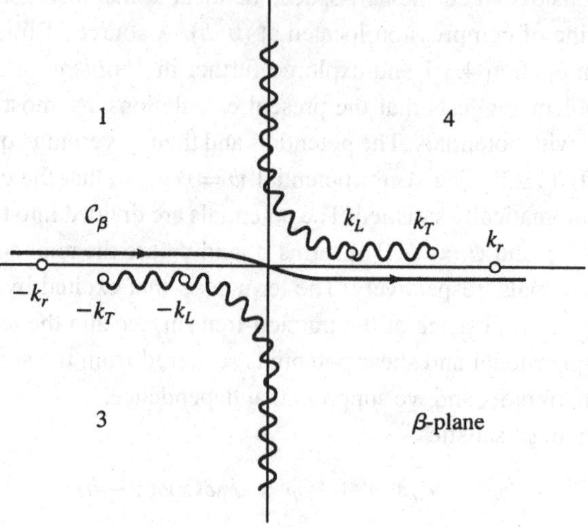

Fig. 5.3. The complex β plane cut so that $\Im(\gamma_I) \geq 0 \,\forall\, \beta$, where $I = L,\ T$. In quadrants 1 and 2, $\Re(\gamma_I) > 0$, and in quadrants 3 and 4, $\Re(\gamma_I) < 0$. The contour C_β is shown, as are the Rayleigh poles at $\pm k_r$.

where

$$\hat{\boldsymbol{p}}_0(\alpha) = \cos\alpha\ \hat{\boldsymbol{e}}_1 - \sin\alpha\ \hat{\boldsymbol{e}}_2. \tag{5.21}$$

In writing (5.20), we have used a notation very similar to that of Section 3.1. Here (5.21) is identical to (3.2) with $\theta_0 = \pi/2 + \alpha$, though we now use α as the independent variable. The α plane, with the contour C, is shown in Fig. 5.4. We discuss its topography further in Section 5.4.

The integrand of (5.20) is a plane, compressional wave incident to the traction-free surface $x_2 = 0$. Using as the integrands the reflected plane, compressional and shear waves calculated in Section 3.1, we can construct the scattered wavefields such that the boundary condition is satisfied, as well as the condition that the scattered wavefields be outgoing. The total scattered wavefield $\boldsymbol{u} = \boldsymbol{u}^{sL} + \boldsymbol{u}^{sT}$, where the first term represents the scattered compressional wavefield and the second the scattered shear wavefield. These wavefields are given by

$$\boldsymbol{u}^{sL} = -\frac{k_L F_0}{4\pi} \int_C \hat{\boldsymbol{p}}_1(\alpha) R_L(\alpha) e^{ik_L \hat{\boldsymbol{p}}_1 \cdot \boldsymbol{x}} e^{ik_L h \sin\alpha}\, d\alpha, \tag{5.22}$$

$$\boldsymbol{u}^{sT} = -\frac{k_L F_0}{4\pi} \int_C \hat{\boldsymbol{d}}_2(\alpha) R_T(\alpha) e^{ik_T \hat{\boldsymbol{p}}_2 \cdot \boldsymbol{x}} e^{ik_L h \sin\alpha}\, d\alpha. \tag{5.23}$$

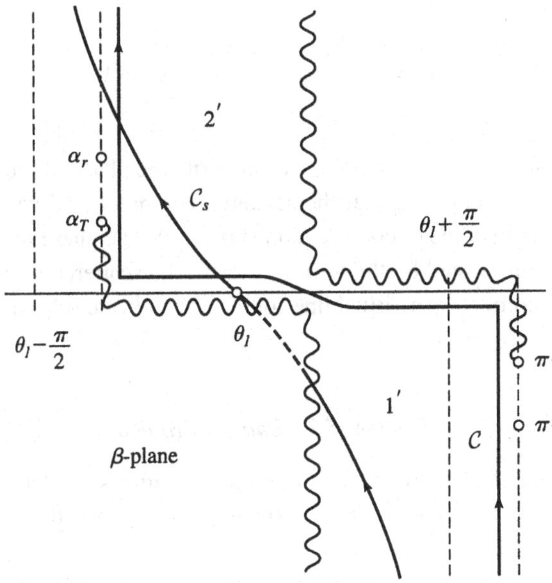

Fig. 5.4. The complex α plane. The branch cuts with branch points α_T and $\pi - \alpha_T$, the two Rayleigh poles α_r and $\pi - \alpha_r$, and the contour C are indicated. Moreover, the saddle point θ_1, for the compressional wave, and the contour of steepest descents C_s, with its asymptotes at $\theta_1 \pm \pi/2$, are also indicated. Quadrants 1 and 2 of Fig. 5.3 are mapped into quadrants 1′ and 2′. Note how part of the contour of steepest descents passes onto the other Riemann sheet. This is indicated by the dashed portion of the contour.

The several unit vectors are given by

$$\hat{\boldsymbol{p}}_1(\alpha) = \cos\alpha\, \hat{\boldsymbol{e}}_1 + \sin\alpha\, \hat{\boldsymbol{e}}_2, \tag{5.24}$$

$$\hat{\boldsymbol{p}}_2(\alpha) = \cos\bar{\alpha}\, \hat{\boldsymbol{e}}_1 + \sin\bar{\alpha}\, \hat{\boldsymbol{e}}_2, \tag{5.25}$$

and

$$\hat{\boldsymbol{d}}_2(\alpha) = \hat{\boldsymbol{e}}_3 \wedge \hat{\boldsymbol{p}}_2(\alpha). \tag{5.26}$$

These are identical to the unit vectors defined in Section 3.1 by (3.4), (3.6) and (3.7), when $\theta_0 = \pi/2 + \alpha$ and $\theta_2 = \pi/2 + \bar{\alpha}$. The two reflection coefficients $R_L(\alpha)$ and $R_T(\alpha)$ are also given by (3.10)–(3.12). The independent variable is taken as α, with the understanding that $c_L^{-1} \cos\alpha = c_T^{-1} \cos\bar{\alpha}$ is used to relate α to $\bar{\alpha}$. Expressing the reflection coefficients in terms of α and $\bar{\alpha}$, we have

$$R_L(\alpha) = A_-(\alpha)/A_+(\alpha), \tag{5.27}$$

$$R_T(\alpha) = 2\kappa \sin 2\alpha \cos 2\bar{\alpha}/A_+(\alpha), \tag{5.28}$$

with

$$A_{\mp} = \sin 2\alpha \sin 2\bar{\alpha} \mp \kappa^2 \cos^2 2\bar{\alpha} \tag{5.29}$$

and $\kappa = c_L/c_T$.

Defining $\boldsymbol{X} = x_1\hat{\mathbf{e}}_1 + (x_2 + h)\hat{\mathbf{e}}_2$, we can write the phase of the integrand of (5.22) as $e^{ik_L\hat{\boldsymbol{p}}_1\cdot\boldsymbol{X}}$, implying that the scattered compressional wave appears to come from a virtual line of compression at $(0, -h)$. The phase of the integrand of (5.23) is complicated by the term $e^{ik_L h \sin\alpha}$. The scattered shear wave does not appear to come from a virtual line source. We indicate subsequently that its virtual source is a caustic.

Problem 5.2 Lamb's Problem

The problem of a buried line of compression is often solved differently from the method we have just used. The present problem indicates this more common method.

Continue to work with the potentials. To apply the boundary conditions, the reader will need τ_{12} and τ_{22} expressed in terms of the potentials. Accordingly, show that

$$\tau_{12} = \mu(2\partial_1\partial_2\varphi + \partial_2\partial_2\psi - \partial_1\partial_1\psi), \tag{5.30}$$

$$\tau_{22} = \lambda\partial_\alpha\partial_\alpha\varphi + 2\mu(\partial_2\partial_2\varphi - \partial_2\partial_1\psi). \tag{5.31}$$

Take a Fourier transform in x_1 of (5.17) and solve the ordinary differential equation in x_2. Note that $[d_2\,{}^*\varphi^i]_{h-}^{h+} = F_0$. Thus show that

$${}^*\varphi^i(\beta, x_2) = -\frac{iF_0}{2\gamma_L}e^{i\gamma_L|x_2-h|}. \tag{5.32}$$

Show that the transformed, scattered potentials are given by ${}^*\varphi = \Phi(\beta)e^{i\gamma_L x_2}$ and ${}^*\psi = \Psi(\beta)e^{i\gamma_T x_2}$. Are the radicals γ_I identical to those defined previously?

Expressing the traction terms as $\tau_{ij}^t = \tau_{ij}^i + \tau_{ij}$, the boundary conditions at $x_2 = 0$ become $\tau_{12} = -\tau_{12}^i$ and $\tau_{22} = -\tau_{22}^i$. Enforce the boundary conditions in the transform domain to show that

$$\Phi(\beta) = \frac{iF_0}{2\gamma_L}e^{i\gamma_L h}\frac{\left[(k_T^2 - 2\beta^2)^2 - 4\beta^2\gamma_L\gamma_T\right]}{R(\beta)}, \tag{5.33}$$

and find a similar expression for $\Psi(\beta)$. The function $R(\beta)$ is the Rayleigh function, (3.47), multiplied by ω^4 to account for the change in scaling.

Lastly, calculate the x_1 component of the scattered particle displacement, $u_1 = u_1^{sL} + u_1^{sT}$. Show that

$$u_1^{sT} = \frac{F_0}{2\pi} \int_{C_\beta} U^{sT}(\beta) e^{i(\beta x_1 + \gamma_T x_2)} e^{i\gamma_L h} \, d\beta, \tag{5.34}$$

where

$$U^{sT} = \frac{2\beta\gamma_T\left(k_T^2 - 2\beta^2\right)}{R(\beta)}. \tag{5.35}$$

Find the expression for u_1^{sL}.

5.3 Asymptotic Approximation of Integrals

The calculations of the previous section, as well as those undertaken in the previous chapters, have indicted how readily one arrives at integrals such as (5.22) or (5.23). We now consider their asymptotic approximation when $k_I r \to \infty$; that is, when the distance r is many wavelengths long. Asymptotic approximations give a satisfying interpretation to these integrals as ray fields. This is a another way of generating asymptotic approximations such as those discussed in Section 2.4. We are then concerned with integrals of the form

$$I(\kappa) = \int_{C_1} f(z) e^{\kappa q(z)} \, dz, \tag{5.36}$$

where $q(z) = u(x_1, x_2) + iv(x_1, x_2)$. Also note that $I(\kappa)$ is also dependent on the contour of integration C_1, though that dependence is seldom explicitly indicated. The parameter[3] κ, which is often $k_I r$, is assumed to be real, positive and large, though the approximations can (with care) be analytically continued to sectors of the complex κ plane.

5.3.1 Watson's Lemma

To facilitate our work we need to study briefly the gamma function and two of its friends. We define the gamma function[4] as

$$(z)! := \int_0^\infty e^{-t} t^z \, dt, \tag{5.37}$$

[3] The first step in asymptotically approximating an integral is to scale the variables so that the large parameter κ can be clearly identified. This scaling and identification depends a good deal on the situation of interest. Harris (1987) indicates how these decisions can effect approximating diffraction from an aperture, while Carrier et al. (1983) indicate how they affect the asymptotic approximation of a Hankel function.

[4] $(z)!$ is also written as $\Gamma(z+1)$.

the incomplete gamma function as

$$\gamma(z, x) := \int_0^x e^{-t} t^{z-1}\, dt, \tag{5.38}$$

and the complement to the incomplete gamma function as

$$\begin{aligned} \Gamma(z, x) &:= (z-1)! - \gamma(z, x), \\ &= \int_x^\infty e^{-t} t^{z-1}\, dt. \end{aligned} \tag{5.39}$$

These definitions are for $\Re(z) > 0$, though the functions can be analytically continued to $\Re(z) < 0$. We take x as real and positive.

Lemma.

$$\Gamma(a, x) \sim e^{-x} \sum_{n \geq 1} \frac{(a-1)!}{(a-n)!} x^{a-n}, \quad x \to \infty, \tag{5.40}$$

where a and x are real and positive.

Proof The following proof is from Copson (1971), but it is repeated here both for completeness and because it illustrates a very simple but useful way to generate an asymptotic approximation to an integral, namely integration by parts. After N such integrations (5.39) becomes

$$\Gamma(a, x) = e^{-x} \sum_{n=1}^{N} \frac{(a-1)!}{(a-n)!} x^{a-n} + \frac{(a-1)!}{(a-N-1)!} \Gamma(a-N, x). \tag{5.41}$$

To estimate the magnitude of the remainder term, we note that

$$\begin{aligned} &\left| \frac{(a-1)!}{(a-N-1)!} \int_x^\infty e^{-t} t^{a-N-1}\, dt \right| \\ &< \left| \frac{(a-1)!}{(a-N-1)!} \right| x^{a-N-1} \int_x^\infty e^{-t}\, dt \\ &= \left| \frac{(a-1)!}{(a-N-1)!} \right| x^{a-N-1} e^{-x}, \end{aligned} \tag{5.42}$$

for $N > (a-1)$. Equation (5.40) follows. □

We begin by considering (5.36) with a contour C_1 that begins at a finite point z_1 and extends to ∞ in a sector of the complex plane wherein the integral is convergent. We assume that $d_z q \neq 0$ even at the end point z_1. We look for a contour C_2 from z_1 to ∞ along which $\Re q(z) < \Re q(z_1)$ and $\Im q(z) = \Im q(z_1)$.

First, we map the z plane into the s plane by using

$$s = q(z_1) - q(z). \tag{5.43}$$

Second, we deform the contour in the s plane to one along the real s axis from zero to ∞. As with the Cagniard–deHoop contour, one seldom works directly in the s plane. Rather it is the contour C_1 in the z plane that is deformed to a new contour C_2 in that plane. Any poles or branch cuts encountered during the deformation must be appropriately surrounded and their contributions added to the asymptotic approximation of the integral, though we do not include such contributions here.

With the change of variables of (5.43) we introduce the function $G(s) = -f(z)/d_z q$ and write $G(s)$ as $G(s) = g(s)s^\lambda$, where $g(s)$ is analytic at $s = 0$ $(z = z_1)$ with a radius of convergence $\bar{r}$. The term s^λ includes any singularity at $s = 0$ either from $f(z)$ itself or as a result of mapping to the s plane. The integral now assumes the form $\exp[\kappa q(z_1)]J(\kappa)$, where $J(\kappa)$ is given by (5.44).

Proposition 5.1 (Watson's Lemma). *Consider the integral*

$$J(\kappa) = \int_0^\infty g(s)s^\lambda e^{-\kappa s}\, ds, \tag{5.44}$$

arrived at by the change of variables described in the preceding paragraphs. $\lambda > -1$. Let real constants K and b exist so that $|g(s)| \le Ke^{bs}$ as $s \to \infty$. Let $g(s)$ be analytic at $s = 0$ so that for $s \in [0, r^*]$,

$$g(s) = a_0 + a_1 s + a_2 s^2 + \cdots + R_{m+1}(s), \tag{5.45}$$

with $|R_{m+1}(s)| \le Cs^{m+1}$; $\bar{r}$ is the radius of convergence for $g(s)$ and $r^* < \bar{r}$. Then

$$J(\kappa) \sim \sum_{n\ge 0} a_n \frac{(\lambda+n)!}{\kappa^{\lambda+n+1}}, \qquad \kappa \to \infty. \tag{5.46}$$

Proof This is proven in Copson (1971), Carrier et al. (1983), Ablowitz and Fokas (1997) and Wong (1989) with greater generality and weaker assumptions than those stated here.

The integral $J(\kappa)$ is written as

$$J = \int_0^{r^*} e^{-\kappa s} s^\lambda (a_0 + a_1 s + a_2 s^2 + \cdots + a_m s^m)\, ds$$
$$+ \int_0^{r^*} e^{-\kappa s} s^\lambda R_{m+1}(s)\, ds + \int_{r^*}^\infty g(s)s^\lambda e^{-\kappa s}\, ds. \tag{5.47}$$

The first integral is again split into two parts by being written as one from zero to ∞ minus one from r^* to ∞. The integral from zero to ∞ is readily evaluated, giving

$$a_0 \frac{\lambda!}{\kappa^{\lambda+1}} + a_1 \frac{(\lambda+1)!}{\kappa^{\lambda+2}} + \cdots + a_m \frac{(\lambda+m)!}{\kappa^{\lambda+m+1}}. \tag{5.48}$$

Each term of the second integral, say that with index n, is such that

$$\int_{r^*}^{\infty} e^{-\kappa s} a_n s^{\lambda+n} \, ds = \frac{a_n}{\kappa^{\lambda+n+1}} \int_{\kappa r^*}^{\infty} e^{-t} t^{\lambda+n} \, dt. \tag{5.49}$$

Noting that the integral on right-hand side is $\Gamma(\lambda+n+1, \kappa r^*)$, we conclude from the previous *Lemma* that each term of this second integral is

$$\int_{r^*}^{\infty} e^{-\kappa s} a_n s^{\lambda+n} \, ds = \mathrm{O}\left(\frac{e^{-\kappa r^*}}{\kappa}\right). \tag{5.50}$$

With $|R_{m+1}(s)| \leq C s^{m+1}$, the second integral in (5.47) is such that

$$\begin{aligned} \left| \int_0^{r^*} e^{-\kappa s} s^{\lambda} R_{m+1}(s) \, ds \right| & \\ \leq \frac{C}{\kappa^{\lambda+m+2}} \int_0^{\kappa r^*} e^{-t} t^{\lambda+m+1} \, dt & \\ = \mathrm{O}\left[\kappa^{-(\lambda+m+2)}\right], & \end{aligned} \tag{5.51}$$

where we recognize that the integral on the right is, after a change of variables, the incomplete gamma function, (5.38). With the use of (5.39) and the previous *Lemma*, the ordering follows. Lastly, using $|g(s)| \leq K e^{bs}$, as $s \to \infty$, and the previous *Lemma*, we find that

$$K \int_{r^*}^{\infty} e^{-s(\kappa-b)} s^{\lambda} \, ds = \mathrm{O}\left[\frac{e^{-r^*(\kappa-b)}}{(\kappa-b)}\right]. \tag{5.52}$$

All the the bounds are taken as $\kappa \to \infty$. The asymptotic approximation (5.46) follows. □

One simple weakening of the assumptions of this proposition follows immediately by noting that $g(s)$ need only have the asymptotic behavior exhibited by (5.45) as $s \to 0$. This fact and *Watson's lemma* are the starting point for the various Abelian and Tauberian theorems found in Doetsch (1974) or van der Pol and Bremmer (1950). It also demonstrates the principle that how a function behaves at its origin is projected into how its transform behaves at infinity and vice versa.

5.3.2 Method of Steepest Descents

We next consider (5.36) with a contour C_1 that stretches across the complex plane, beginning in one sector of the complex plane at infinity and ending in another sector at infinity. The contours C_β and C defining integrals (5.19) and (5.20) are examples. Assume that $\Re(q)$ has a stationary point at z_s. We then deform the contour C_1 into a contour C_s passing through this point and defined by $\Re q(z) \leq \Re q(z_s)$ and $\Im q(z) = \Im q(z_s)$, assuming, of course, that such a deformation is possible. Pole and branch cut contributions arising from deforming C_1 to C_s must be added to the contribution made by the new integral over C_s, though they are not explicitly included in the present discussion. If $f(z)$ is analytic and slowly varying near z_s, the principal contribution to the new integral comes from z_s and $I(\kappa)$ takes the approximate form

$$I(\kappa) \approx f(z_s) e^{i\kappa v(x_{1s}, x_{2s})} \int_{C_s} e^{\kappa u(x_1, x_2)} dz. \tag{5.53}$$

The Steepest Descents Contour

To fix our ideas more precisely, we examine the topography of the function $q(z) = u(x_1, x_2) + iv(x_1, x_2)$ near its stationary point z_s. Our discussion follows a similar one given in Felsen and Marcuvitz (1994). We assume that $q(z)$ is analytic in region $\mathcal{Q}$ containing z_s, and that $d_z q = 0$ but $d_z^2 q \neq 0$ at z_s. The function $q(z)$ satisfies the Cauchy–Riemann equations

$$\partial_1 u = \partial_2 v, \qquad \partial_2 u = -\partial_1 v \tag{5.54}$$

in $\mathcal{Q}$ so that u and v are harmonic functions. Thus, if at (x_{1s}, x_{2s}) the curvature of the surface $u(x_1, x_2) = C$, or $v(x_1, x_2) = C$, where C is a constant, is positive along the x_1 axis, it is negative along the (perpendicular) x_2 axis. The stationary point z_s is therefore a saddle point and its neighborhood a col or saddle (Courant and John, 1974). Figure 5.5 sketches the topography in the neighborhood of z_s.

To investigate this neighborhood and the contour C_s, we parameterize the contour C_s with the arclength r so that the directional derivative of u along this contour is

$$d_r u = \partial_1 u \cos\alpha + \partial_2 u \sin\alpha, \tag{5.55}$$

where $d_r x_1 = \cos\alpha$ and $d_r x_2 = \sin\alpha$. Viewing (5.55) as a function of α, the direction of maximum change in u is given by

$$d(d_r u)/d\alpha = -\partial_1 u \sin\alpha + \partial_2 u \cos\alpha = 0. \tag{5.56}$$

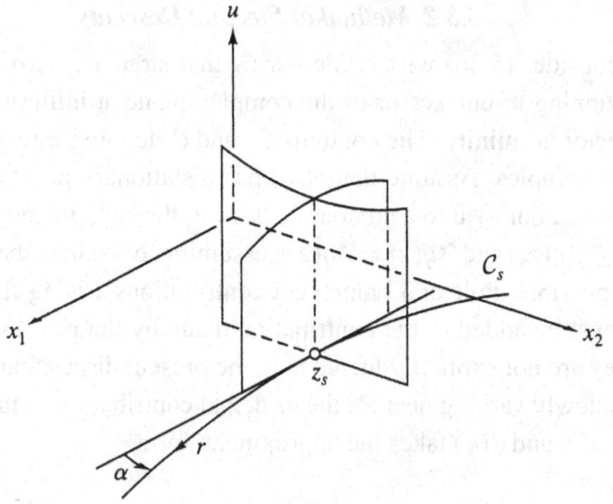

Fig. 5.5. A three-dimensional sketch of the topography of $\Re(q) = u$ near z_s, the stationary point. The contour C_s, its arclength r, and the angle it makes with the x_1 axis as it passes through z_s are shown.

With the use of the Cauchy–Riemann equations, this equation becomes $d_r v = 0$. Therefore v is constant along the same contour on which u changes most rapidly. However, there are two such contours and we need to select that one along which u decreases, as we move away from the saddle point z_s. The contour C_s along which v is constant but along which u decreases most rapidly is called the *path of steepest descents*, or the *steepest descents contour*.

To determine this contour, we examine the behavior of $q(z)$ in the neighborhood of z_s. Expanding $q(z)$ about z_s gives

$$\exp[\kappa\, q(z)] \approx \exp[\kappa\, q(z_s)] \exp\left[(\kappa/2)\, d_z^2 q(z_s)(z - z_s)^2\right], \tag{5.57}$$

or

$$\exp[\kappa\, q(z)] \approx \exp[\kappa\, q(z_s)] \times \exp\left[(\kappa/2)\left| d_z^2 q(z_s)(z - z_s)^2\right| (\cos 2\psi + i \sin 2\psi)\right], \tag{5.58}$$

where

$$\psi = \arg(z - z_s) + \tfrac{1}{2} \arg\left[d_z^2 q(z_s)\right]. \tag{5.59}$$

Examining (5.58) indicates that, locally, $e^{\kappa u}$ decreases most rapidly and $e^{i\kappa v}$ remains constant for $\psi = \pm\pi/2$. This then is the contour C_s we seek. For $\psi = 0$ or π, $e^{\kappa u}$ increases most rapidly and $e^{i\kappa v}$ again remains constant; for $\psi = \pm\pi/4$ or $\pm 3\pi/4$, $e^{\kappa u}$ remains constant and $e^{i\kappa v}$ oscillates most rapidly. These latter contours are *paths of stationary phase* or *stationary phase contours*.

Note that we could shift the steepest descents contour to pass through a point higher up the ridge. However, this would mean that the exponential in the integrand would oscillate and, therefore, there could be cancellations, voiding the assumption that the maximum contribution to the integral comes from this point.

An Isolated First-Order Saddle Point

With the following proposition we put the knowledge just gained into a more precise form. As with previous propositions, the conditions cited are stronger than they have to be. Any of the references cited when *Watson's lemma* was discussed also discuss the method of steepest descents from various points of view.

Proposition 5.2. *Let the function $q(z) = u(x_1, x_2) + iv(x_1, x_2)$ be analytic in a region $\mathcal{Q}$ containing z_s. The point z_s is a saddle point, and it is isolated and first order. That is, $d_z(z_s) = 0$, but $d_z^n q(z_s) \neq 0,\ n \geq 2$.*

Assume that the contour C_1 has been deformed into the steepest descents contour C_s defined previously.

Let the function $f(z)$ be analytic in a region $\mathcal{R}$ containing z_s such that $f(z_s) \neq 0$. Moreover, let there be constants K and b such that

$$\left| \frac{2\, f(z)[q(z_s) - q(z)]^{1/2}}{d_z q(z)} \right| < K e^{b[q(z)-q(z_s)]},$$

as $|z| \to \infty$ in sectors of the complex plane where C_s begins and ends.

Then

$$I(\kappa) \sim \frac{(-2\pi)^{1/2}}{\left[\kappa\, d_z^2 q(z_s)\right]^{1/2}} f(z_s) e^{i\kappa q(z_s)} + O\left(\kappa^{-3/2}\right), \qquad \kappa \to \infty. \qquad (5.60)$$

The argument of $[-2/d_z^2 q(z_s)]^{1/2}$ is defined by the argument of $d_s z$ at z_s.

Proof We again use a mapping that captures the essential topographical features of the phase function $q(z)$ in the neighborhood of the stationary point. We map from the z plane to the s plane by using

$$s^2 = q(z_s) - q(z). \qquad (5.61)$$

We define the inverse mapping so that, as s varies from $-\infty$ to ∞, z traces the steepest descents contour from beginning to end. Figure 5.5 suggests how the contour C_s might look in the z plane. With this change of integration variable, the integral $I(\kappa)$ is now given by

$$I(\kappa) = e^{\kappa q(z_s)} \int_{-\infty}^{\infty} G(s) e^{-\kappa s^2}\, ds, \qquad (5.62)$$

where

$$G(s) = f(z)\frac{dz}{ds}, \qquad \frac{dz}{ds} = \frac{-2s}{d_z q(z)}. \tag{5.63}$$

Note that at $z = z_s$, $d_s z$ becomes indefinite so that a limit must be taken as $z \to z_s$.

We have assumed that $f(z)$ is analytic in a region containing z_s. Moreover, $d_s z$ is also analytic in a region containing $s = 0$ (the singularity is removable). Thus $G(s)$ is analytic in a region containing $s = 0$ and can be expanded in a power series with a radius of convergence $\bar{r}$.

$$G(s) = G(0) + d_s G(0)s + d_s^2 G(0)(s^2/2) + \cdots + R_{2(m+1)}(s), \tag{5.64}$$

where $|s| \leq r^* < \bar{r}$. This expansion is seldom easy to calculate because (5.61) expanded in a power series in $z - z_s$ must be inverted to give $z = z(s)$ as a power series in s. Copson (1935) describes how to invert a series. However, it is usually enough to calculate only the first term $G(0)$[5]. Therefore

$$G(0) = f(z_s)\left(\frac{dz}{ds}\right)_{s=0}, \quad \left(\frac{dz}{ds}\right)_{s=0} = \left[\frac{-2}{d_z^2 q(z_s)}\right]^{1/2}. \tag{5.65}$$

The differential element ds is real and positive along the path of integration, so that at $s = 0$,

$$\arg\left(\frac{dz}{ds}\right)_{s=0} = \arg(dz)_{z=z_s} = \alpha. \tag{5.66}$$

Because $G(s)$ is analytic, $|R_{2(m+1)}(s)| < Cs^{2(m+1)}$, where C is a constant. Moreover, setting $G(s) = g(s)s$, we note that by hypothesis $|g(s)| < Ke^{bs^2}$ as $|s| \to \infty$. At this point we follow a procedure almost identical to that used to establish *Proposition 5.1.* Setting $I(\kappa) = \exp[\kappa q(z_s)]J(\kappa)$, we write

$$J = \int_{-r^*}^{r^*} e^{-\kappa s^2}\left[G(0) + \cdots + d_s^{2m} G(0)\frac{s^{2m}}{(2m)!}\right] ds$$
$$+ 2\int_0^{r^*} e^{-\kappa s^2} R_{2(m+1)}(s)\, ds + \int_{r^*}^{\infty} g(s) s\, e^{-\kappa s^2}\, ds. \tag{5.67}$$

Estimating the contributions of the various integrals is identical to that used previously in (5.48)–(5.52). The asymptotic approximation, (5.60), follows. □

Note how r^* enters the calculation. If r^* is small we have not really achieved very much, so that we want it to be at least O(1). This is what is meant by the phrase that $f(z)$ be slowly varying near z_s.

[5] If, say, $d_s^2 G(0)$ is the first nonzero term, the pattern of the proof is unchanged.

5.3.3 Stationary Phase Approximation

The stationary phase approximation is usually applied to integrals of the form

$$I(\kappa) = \int_{x_1}^{x_2} f(x)e^{i\kappa p(x)}\,dx, \qquad x_1 < x_2, \tag{5.68}$$

where $p(x)$ is a real function when x is real, and κ is real, positive, and assumed large. The contour is thus a stationary phase contour. We assume that there is a first-order stationary point x_s between x_1 and x_2.

To asymptotically approximate this integral, we use the method of steepest descents. We assume $ip(z)$ is analytic in region containing x_s so that the stationary point becomes a saddle point. The integration contour is then deformed to a path of steepest descents passing through x_s. Using (5.60), we find the contribution from x_s is

$$I(\kappa) \sim \left[\frac{2\pi}{\kappa\left|d_x^2 p(x_s)\right|}\right]^{1/2} f(x_s)e^{i[\kappa p(x_s)\pm\pi/4]}, \qquad \kappa \to \infty. \tag{5.69}$$

The plus sign is taken for $d_x^2 p(x_s) > 0$ and the minus sign for $d_x^2 p(x_s) < 0$. The end point contributions are estimated by using the method outlined in Section 5.3.1 and *Watson's lemma.* The contributions from the end points x_1 and x_2 are

$$I(\kappa) \sim \frac{1}{i\kappa}\left[\frac{f(x_2)}{d_x p(x_2)}e^{i\kappa p(x_2)} - \frac{f(x_1)}{d_x p(x_1)}e^{i\kappa p(x_1)}\right], \qquad \kappa \to \infty. \tag{5.70}$$

The integral (5.68) is asymptotically approximated by the sum of the contributions given by (5.69) and (5.70).

The references cited in connection with *Watson's lemma* discuss the stationary phase approximation from various other viewpoints and provide more detailed discussions.

Problem 5.3 Asymptotic Approximations of Integrals

Problem 1. Consider an integral having the form given by (5.7). Show that the Cagniard–deHoop contour is one of steepest descents and that $\alpha = -s\cos\theta$ is a saddle point. The angle θ is that shown in Fig. 5.2.

Problem 2. Consider the integral given by (5.13). Use Watson's lemma to show that the first term of an asymptotic expansion in p is

$$\bar{I}(x_1, x_2, p) \sim \frac{a_0(-1/2)!}{p^{1/2}}e^{-psr}, \qquad p \to \infty. \tag{5.71}$$

Find a_0. Invert this to approximate $I(x_1, x_2, t)$. This approximation is accurate as $t \to sr$ (consult Doestch, 1974 or van der Pol and Bremmer, 1950 to learn why) and is therefore sometimes called a wavefront approximation.

Problem 3. Consider an integral of the form

$$I(kr) = \int_C P(\cos\alpha) e^{ikr\cos(\theta-\alpha)}\, d\alpha, \tag{5.72}$$

where $\theta \in (0, \pi)$. The contour C begins at $\pi - i\infty$ and ends at $i\infty$. Sketch the contour and show that the integral converges provided it begins and ends in the sectors containing these points. Speculate on the structure that $P(\cos\alpha)$ might be permitted in order that an asymptotic approximation of the integral be successful. Consider the change of integration variable

$$\tau = 2^{1/2} e^{i\pi/4} \sin[(\theta - \alpha)/2]. \tag{5.73}$$

Show that a deformation of the contour to one along which τ is real gives an integral along the path of steepest descents C_s.

5.4 Buried Harmonic Line of Compression II

We now turn to the asymptotic approximation of the integrals, (5.22) and (5.23), that describe the wavefield scattered from the traction-free surface.

5.4.1 The Complex Plane

We begin by recalling the Sommerfeld transformation used to arrive at (5.20), namely

$$\beta = k_L \cos\alpha = k_T \cos\bar{\alpha}, \quad \gamma_L = k_L \sin\alpha, \quad \gamma_T = k_T \sin\bar{\alpha}, \tag{5.74}$$

where $c_L^{-1} \cos\alpha = c_T^{-1} \cos\bar{\alpha}$ and $\gamma_I = (k_I^2 - \beta^2)^{1/2}$. Moreover, recall Figs. 5.3 and 5.4 where the features of the β plane and the α plane, respectively, are described. In particular, the contours C_β and C are indicated there. Section 3.4.3 points out that $A_+(\alpha)$ is identical to the Rayleigh function, provided the definition of α given in the paragraph preceding (5.27) is used. The integrands of (5.22) and (5.23) therefore have poles at α_r and $\pi - \alpha_r$. These are indicated in Fig. 5.4, as are their β-plane counterparts in Fig. 5.3. These poles give rise to oppositely directed Rayleigh surface waves.

Before continuing, it is useful to ask what Fig. 5.4 and (5.22) and (5.23) mean. Consider the values of α to the left of the vertical line through $\pi/2$.

Those to the right have a very similar meaning. From $\pi/2$ to zero the incident compressional wave is composed of plane waves, each incident at the angle α in this range. They are reflected into plane compressional and shear waves propagating away from the surface at α and $\bar{\alpha}$, respectively. As α climbs the imaginary axis, incident inhomogeneous plane waves are added to the overall incident wavefield. They decay toward the surface. Reciprocally they scatter into compressional inhomogeneous plane waves that decay away from the surface. However, as α takes on imaginary values from zero to a critical angle α_T, defined by $c_L^{-1}\cos\alpha_T = c_T^{-1}$, shear plane waves are still being reflected at the real angle $\bar{\alpha}$. At the limiting (imaginary) angle α_T a shear wave grazing the surface is excited. For values of α beyond α_T the scattered shear plane waves are now also inhomogeneous and decay away from the surface. Lastly, note that as we climb beyond α_T, we quite quickly encounter the Rayleigh pole at α_r. This term must be included in our evaluation of (5.22) and (5.23) when it is enclosed. Note that the Rayleigh wave could not be excited unless the incident disturbance contained inhomogeneous waves.

5.4.2 The Scattered Compressional Wave

Setting $x_1 = r\cos\theta_1$ and $(x_2 + h) = r\sin\theta_1$, we express (5.22) as

$$u^{sL} = -\frac{k_L F_0}{4\pi}\int_C \hat{p}_1(\alpha)R_L(\alpha)e^{ik_L r\cos(\alpha-\theta_1)}\,d\alpha. \tag{5.75}$$

To approximate this integral for $k_L r$ large, we must first ascertain where the contour of steepest descents C_s begins and ends, and what it looks like near the stationary point. Expressing α as $\alpha = \alpha_1 + i\alpha_2$,

$$\cos(\alpha-\theta_1) = \cos(\alpha_1-\theta_1)\cosh\alpha_2 - i\sin(\alpha_1-\theta_1)\sinh\alpha_2. \tag{5.76}$$

For the integral (5.75) to converge, $\Im[\cos(\theta_1-\alpha)] > 0$ so that C_s must begin and end in a region of the α plane where $\sin(\alpha_1-\theta_1)\sinh\alpha_2 < 0$. That is, the contour must begin in a region $\alpha_1 \in (\theta_1, \pi+\theta_1)$, where $\alpha_2 < 0$, and end in $\alpha_1 \in (\theta_1-\pi, \theta_1)$, where $\alpha_2 > 0$.

The function $q(\alpha) = i\cos(\alpha-\theta_1)$ and the stationary point, a saddle point, is given by $\alpha = \theta_1$. The contour C_s is described by $\Im[q(\alpha)] = 1$, or, from (5.76), by

$$\cos(\alpha_1-\theta_1)\cosh\alpha_2 = 1. \tag{5.77}$$

Near $\alpha = \theta_1$, C_s is described by $\alpha_2 = \pm(\alpha_1-\theta_1)$. Because we want $\Re[q(\alpha)]$ to achieve a maximum along C_s at θ_1, $\alpha_2 = -(\alpha_1-\theta_1)$. Further, for $|\alpha|$ large, $\alpha_1 \to \pm\pi/2+\theta_1$.

Figure 5.4 indicates the contour C_s passing through the saddle point θ_1, as well as the two limiting lines at $\alpha_1 = \pm\pi/2 + \theta_1$. Note that depending upon θ_1, the poles at α_r and $\pi - \alpha_r$ may or may not be enclosed. *Problem 5.4* asks the reader to evaluate the contribution from α_r. Suppose that θ_1 is such that in distorting C to C_s the pole at α_r is enclosed. From (5.77), the condition for this first to happen is that $\cos\theta_1 \cosh(-i\alpha_r) = 1$. This is equivalent to the condition that $\cos\theta_{1r} = c_r/c_L$, where we have added the subscript r to indicate this special value. Thus for values of $\theta_{1r} < \theta_1 < \pi/2$, no Rayleigh wave is present. If we inverted our result in time we should find that this is equivalent to asserting that the Rayleigh wave cannot be excited before the incident compressional disturbance strikes the surface.

Note also that C_s passes onto the other Riemann sheet – the dashed portion in Fig. 5.4 – over part of its path. One must be careful to ensure that when this happens that one can reemerge onto the Riemann sheet of physical interest. Occasionally one must also be careful that one does not forget to include any pole contributions that may arise from poles on this other sheet, should they be enclosed by this excursion. When one cannot reemerge onto the Riemann sheet of physical interest, then the contour C_s must instead be wrapped around the branch cut, on the physically meaningful Riemann sheet, and this contribution asymptotically approximated. Harris and Pott (1985) discuss such contributions.

The remaining issue before using the steepest descents result (5.60) is to settle what the argument of the square root is. Following the rule set out in *Proposition 5.2*, $\arg[-2/d_\alpha^2 q(\theta_1)]^{1/2} = 3\pi/4$. Putting the various pieces together, we find that

$$\mathbf{u}^{sL} \sim \frac{A}{4\pi k_L}\left(\frac{2\pi}{k_L r}\right)^{1/2} \hat{\mathbf{p}}_1(\theta_0) R_L(\theta_0) e^{ik_L r} e^{-i\pi/4}, \quad k_L r \to \infty, \tag{5.78}$$

where $F_0 = A/k_L^2$ has been used. Note that this has given us a representation of the kind investigated in Section 2.4. The overall structure of (5.78) is that of a farfield expression for a cylindrical wave radiating from a virtual source at $(0, -h)$, in agreement with what we indicated previously. At the surface $x_2 = 0$, $r = h/\sin\theta_1$ is the radius of curvature of the reflected cylindrical wave as it is just about to form.

5.4.3 The Scattered Shear Wave

Equation (5.23), which is repeated in the following equation, is a somewhat harder integral to approximate.

$$\mathbf{u}^{sT} = -\frac{k_L F_0}{4\pi} \int_C \hat{\mathbf{d}}_2(\alpha) R_T(\alpha) e^{ik_T \hat{\mathbf{p}}_2 \cdot \mathbf{x}} e^{ik_L h \sin\alpha}\, d\alpha. \tag{5.79}$$

The complexity arises from the mixing of wave types: an incident compressional wave whose phase lingers in the term $e^{ik_L h \sin\alpha}$ and the scattered shear wave whose phase is $e^{ik_T \hat{p}_2 \cdot \boldsymbol{x}}$. Finding C_s and investigating it in the neighborhood of the stationary point is similar to that for the compressional case. We therefore know approximately what C_s looks like and can apply (5.60) without working out C_s in detail. Instead, the key to understanding (5.79) can be found by considering the geometry of the rays. We know from (5.78) that a steepest descents approximation will give phase terms that are those of a ray theory.

The first clue can be found by examining (2.51) and (2.56). These expressions are singular when the distance along the ray is the negative of the radius of curvature. That is, the radius of curvature has its origin either at a point on a caustic surface or at a single point. We have seen that the reflected compressional wave (5.78) appears to originate at the virtual source point $(0, -h)$. In a similar way we should expect that the reflected shear wave will appear to originate, if not from a virtual source point, then from a point on a virtual caustic. We have sketched this possibility in Fig. 5.6. The radius of curvature ρ_T measures the distance from a point on the virtual caustic to a point on the surface $x_2 = 0$ at which the scattered shear wave is starting to form. Let s_0 be the distance the compressional ray propagates from $(0, h)$ to this same point on the surface and s_2 be the distance the reflected shear ray propagates from this point to the

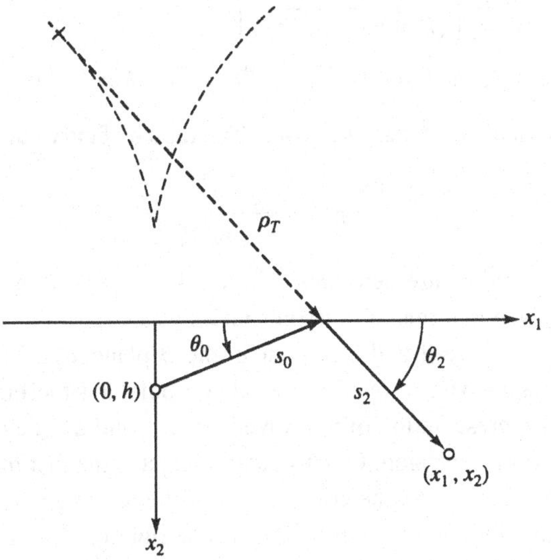

Fig. 5.6. A sketch of the shear ray scattered from the surface when struck by an incident compressional ray. The compressional ray originates at $(0, h)$ and propagates a distance s_0, striking the surface at an angle θ_0. The scattered shear ray emerges at an angle θ_2. It appears to originate from a point on a virtual caustic located by extending the ray backward a distance ρ_T.

observation point (x_1, x_2). We do not at present have an analytic expression for ρ_T, nor do we have one for the virtual caustic. However, we know that the asymptotic expansion of (5.79) must contain the term $(1 + s_2/\rho_T)^{1/2}$ in its denominator so that we can identify ρ_T from the final asymptotic expression.

Setting $(x_1 - s_0 \cos\theta_0) = s_2 \cos\theta_2$ and $x_2 = s_2 \sin\theta_2$, we express (5.79) as

$$\boldsymbol{u}^{sT} = -\frac{k_L F_0}{4\pi} \int_{C_s} \hat{\boldsymbol{d}}_2(\alpha) R_T(\alpha) e^{(k_T s_2 + k_L s_0) q(\alpha)} \, d\alpha, \tag{5.80}$$

where

$$q(\alpha) = i \frac{k_T s_2}{(k_T s_2 + k_L s_0)} \cos(\overline{\alpha} - \theta_2) + i \frac{k_L s_0}{(k_T s_2 + k_L s_0)} \cos(\alpha - \theta_0). \tag{5.81}$$

The parameter $(k_T s_2 + k_L s_0)$ is large. The stationary point, a saddle point, is given by $\overline{\alpha} = \theta_2$ and $\alpha = \theta_0$. This is a manifestation of Fermat's principle that the ray path is an extremum. The unknowns at this point can be taken as s_0, s_2, θ_0, and θ_2, while x_1, x_2 and $h\ (= s_0 \sin\theta_0)$ can be considered as given. Note, as well, that the phase-matching condition $c_L^{-1} \cos\theta_0 = c_T^{-1} \cos\theta_2$ gives a fourth relation. Thus we may solve for the unknowns in terms of what is given.

We have enough information at this point to approximate (5.80) by using (5.60). This gives, after some algebraic manipulation,

$$\boldsymbol{u}^{sT} \sim \frac{A}{4\pi k_L} \left[\frac{2\pi}{(k_L s_0)(1 + s_2/\rho_T)} \right]^{1/2}$$
$$\times\, \hat{\boldsymbol{d}}_2(\theta_0) R_T(\theta_0) e^{i(k_T s_2 + k_L s_0)} e^{-i\pi/4}, \quad k_T s_2 + k_L s_0 \to \infty, \tag{5.82}$$

where again $F_0 = A/k_L^2$ has been used. The radius of curvature ρ_T is given by

$$\rho_T = s_0 \frac{c_L}{c_T} \frac{\sin^2\theta_2}{\sin^2\theta_0}, \tag{5.83}$$

and s_0, s_2, θ_0, and θ_2 are determined in terms of x_1, x_2, h, and the ratio $\kappa = c_L/c_T$. κ is given in terms of Poisson's ratio by (3.13).

In closing we note that the contour in the β plane, C_β, Fig. 5.3, could be distorted so as to wrap around the Rayleigh pole and the branch cuts. This would give a representation of the wavefields $\boldsymbol{u}^{sL}$ and $\boldsymbol{u}^{sT}$ that corresponds to an eigenfunction expansion. Of particular interest is the fact that the spectrum has a discrete eigenvalue, the contribution of the Rayleigh pole, along with a continuous distribution of eigenvalues, the contribution from the branch cuts.

Problem 5.4 The Rayleigh Wave

Calculate the Rayleigh-pole contributions to $\boldsymbol{u}^{sL}$ and $\boldsymbol{u}^{sT}$, and hence the Rayleigh wave excited by a harmonic line of compression at $(0, h)$. There are a number of starting points. I should likely begin with (5.22) and (5.23), but this is not the only starting point.

5.5 Diffraction of an Antiplane Shear Wave at an Edge

The previous problem has considered scattering of a cylindrical wave from an infinite traction-free surface. The complications came about not only because both compressional and shear wavefields were scattered from the surface, but also because the incident wavefield was composed of a complete spectrum of plane waves. In this section we consider a related problem: one in which the scatterer has an edge that excites a full spectrum of plane waves, though the incident wave is a single plane one. Specifically, we consider a time harmonic, antiplane shear wave normally incident to a semi-infinite slit or crack. The crack lies along the positive x_1 axis. On both sides, the traction acting on the surfaces of the crack is zero. Figure 5.7 indicates the geometry of the problem. The equations of motion are given by (1.14) and (1.15), with the corresponding time-harmonic equation being given by (2.20). As we have done previously when dealing with antiplane shear waves, we simplify the notation by dropping

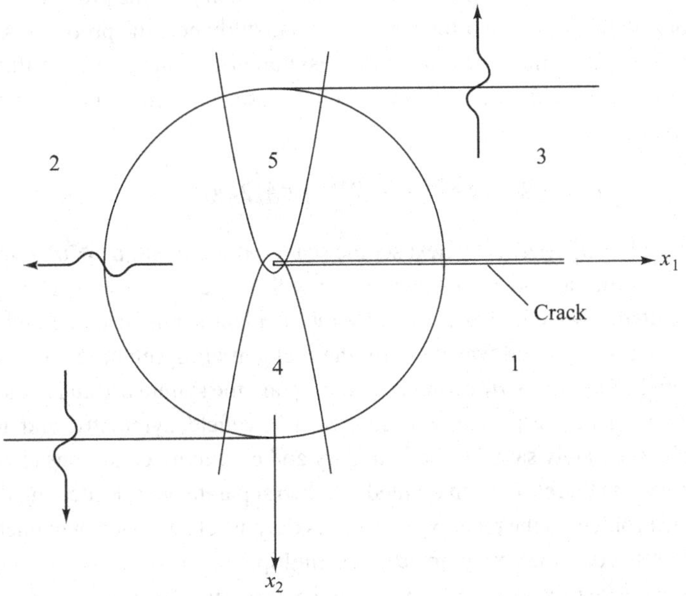

Fig. 5.7. A sketch of the crack, the cylindrical diffracted wave emanating from the crack tip, and the reflected and transmitted geometrical waves. The rippled arrows indicate their directions of propagation. Separating the diffracted and geometrical wavefields, in regions 1,2, and 3, are parabolic shaped, transition, or boundary layers, labeled regions 4 and 5.

the subscript T and use c and $k=\omega/c$ as the wavespeed and wavenumber, respectively.

5.5.1 Formulation

The plane wave u_3^i, described by

$$u_3^i = (A/k)e^{ikx_2}, \tag{5.84}$$

is incident to the crack. The constant A is dimensionless, having been made so by dividing by k, the wavenumber. The total wavefield $u_3^t = u_3^i + u_3$, where u_3 is the scattered wavefield. The boundary and continuity conditions to be satisfied at $x_2 = 0$ are

$$\mu\partial_2 u_3^t(x_1, 0^{\pm}) = 0, x_1 > 0; \quad u_3^t(x_1, 0^+) = u_3^t(x_1, 0^-), \quad x_1 < 0. \tag{5.85}$$

The superscripts $\pm$ indicate that the boundary is approached through positive or negative values of x_2, respectively. Also we ask that the scattered wave be outgoing from its source, the crack, and that it therefore satisfy the principle of limiting absorption (Section 4.4).

Note that $x_2 = 0$ is a plane of reflection symmetry for the problem. As a way to formulate the problem for the scattered disturbance, the problem is divided into two, one symmetric and one antisymmetric with respect to this plane. To begin, we divide the incident wavefield into symmetric and antisymmetric components as follows.

$$u_3^i = (A/2k)(e^{ikx_2} + e^{-ikx_2}) + (A/2k)(e^{ikx_2} - e^{-ikx_2}), \tag{5.86}$$

so that $u_3^i = u_3^{is} + u_3^{ia}$. We next set the scattered wavefield $u_3 = u_3^s + u_3^a$ where u_3^s is symmetric and u_3^a antisymmetric with respect to $x_2 = 0$. The symmetric scattered wavefield is excited by the incident symmetric one and the antisymmetric scattered wavefield by the incident antisymmetric one. The total wavefield $u_3^t = u_3^{ts} + u_3^{ta}$, where u_3^{ts} and u_3^{ta} are the symmetric and antisymmetric components, respectively. Lastly, each wavefield, symmetric and antisymmetric, separately satisfies the boundary and continuity conditions (5.85). The problem has therefore been divided into two separate ones. Reflecting the symmetric problem in the plane $x_2 = 0$ leaves the particle displacement unchanged, while reflecting the antisymmetric one multiplies it by -1.

We consider the symmetric problem. By construction, $\partial_2 u_3^{is} = 0 \,\forall\, x_1$ at $x_2 = 0$. From the symmetry, $\partial_2 u_3^s = 0 \,\forall\, x_1$ at $x_2 = 0$. This and the condition that the scattered wavefield propagate outward imply that $u_3^s \equiv 0$. Therefore the symmetric part of the incident wavefield remains unaffected by the presence of the crack in its path of propagation. The complete problem is then equivalent to the antisymmetric one and the scattered wavefield is antisymmetric in x_2.

Reasserting the decomposition $u_3^t = u_3^i + u_3$, but noting that $u_3 = u_3^a$, we are led to reformulate the problem, in the half-space $x_2 > 0$, for the scattered wavefield u_3 as follows.

$$\partial_\alpha \partial_\alpha u_3 + (k + i\epsilon)^2 u_3 = 0, \tag{5.87}$$

subject to the conditions, at $x_2 = 0$,

$$\partial_2 u_3(x_1, 0) = -iAe^{-\epsilon x_1}, \quad x_1 > 0; \qquad u_3(x_1, 0) = 0, \quad x_1 < 0. \tag{5.88}$$

As well, the condition that u_3 be outgoing and satisfy the principle of limiting absorption is enforced. It is for this latter reason that the $i\epsilon$, $0 < \epsilon \ll 1$, has been explicitly added to k in (5.87). We set $k' = k + i\epsilon$, with k taken as real and positive. The addition of the $e^{-\epsilon x_1}$ in the boundary condition is an artifice also needed to enforce the principle of limiting absorption. In this particular problem, because the incident wave strikes the crack normally, the scattered wavefield will contain geometrically reflected waves that, as with the incident wave, do not vanish as $|kx_1| \to \infty$ unless this artifice is used. However, it is not needed for other angles of incidence, as we indicate in *Problem 5.5.* Lastly, we must also append to these equations an *edge condition* as explained in Section 4.7.2. In fact we use the analysis of Section 4.7.3 to demand that

$$ku_3 = \mathrm{O}\big[(kr)^{1/2}\big], \quad kr \to 0, \tag{5.89}$$

where $r = (x_1^2 + x_2^2)^{1/2}$. This condition is essential to ensure that the solution be unique. The scattered wavefield for $x_2 < 0$ is found by using the fact that $u_3(x_1, x_2) = -u_3(x_1, -x_2)$.

We could solve this problem by formulating an integral equation for the scattered wavefield following the ideas outlined in *Problem 4.3.* We should then find that the integral equation can be solved by using the Wiener–Hopf method (Titchmarsh, 1948). However, it is easier, and perhaps just as informative, to solve the problem directly with the Wiener–Hopf method as used by Jones (Jones, 1952; Noble, 1988).

5.5.2 Wiener–Hopf Solution

We begin by setting

$$u_3(x_1, 0) = \begin{cases} 0, & x_1 < 0, \\ u_{3+}, & x_1 > 0, \end{cases} \tag{5.90}$$

$$\partial_2 u_3(x_1, 0) = \begin{cases} \tau_-, & x_1 < 0, \\ \tau_+ = -iAe^{-\epsilon x_1}, & x_1 > 0. \end{cases} \tag{5.91}$$

Our strategy in solving the problem for the scattered wavefield is to find a functional relation between the two unknowns τ_- and u_{3+}.

We seek a solution to (5.87) by using a representation very similar to that used previously in Section 2.3.1 (the roles of x_1 and x_2 are interchanged). That is, we set

$$u_3 = \frac{1}{2\pi} \int_{-\infty}^{\infty} {}^*u_{3+}\, e^{i(\beta x_1 + \gamma x_2)}\, d\beta, \tag{5.92}$$

where $\gamma = (k'^2 - \beta^2)^{1/2}$. The transform ${}^*u_{3+}$ remains to be determined. Recall that $x_2 > 0$ so that the β plane is cut such that $\Im(\gamma) \geq 0,\ \forall\, \beta$. This is the same choice of Riemann sheet as was taken in Section 5.2 and first explicitly discussed in Section 3.4.4. Figure 5.8 shows the branch points with the accompanying cuts.

Imposing the conditions at $x_2 = 0$ on (5.92) and using (5.90) gives the following:

$${}^*u_{3+} = \int_0^{\infty} u_{3+}(x_1) e^{-i\beta x_1}\, dx_1. \tag{5.93}$$

With some thought, one realizes that, to the right along the crack in Fig. 5.6, the dominant wavefield is a plane one, propagating normally away from the

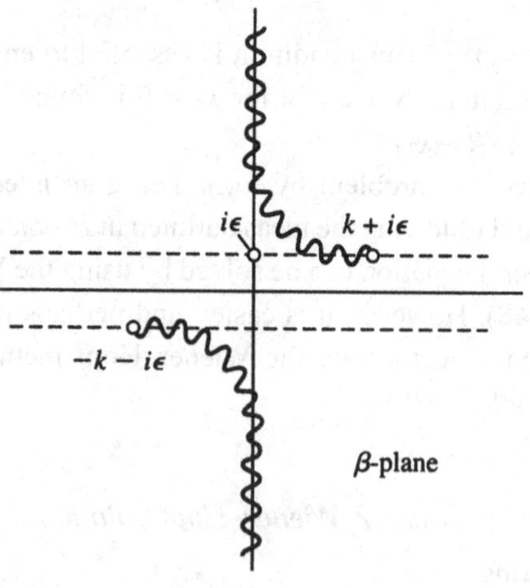

Fig. 5.8. The complex β plane showing the branch cuts, pole at $i\epsilon$, and strip of common analyticity 2ϵ wide. The contour for the inverse transform initially lies within this strip. As $\epsilon \to 0$ the pole will approach a position just above this contour.

crack, but one that is forced to decay as $x_1 \to \infty$ because of the $e^{-\epsilon x_1}$ placed in the first of (5.88). Any other wave present will originate from the crack tip.[6] That part grazing the crack face will also decay as $e^{-\epsilon x_1}$ because $k' = k + i\epsilon$. Thus $ku_{3+} = O(e^{-\epsilon x_1})$ as $kx_1 \to \infty$, from which it follows that ${}^*u_{3+}$ is an analytic function of β for $\Im(\beta) < \epsilon$. Noble (1988) discusses in some detail the conditions needed to ensure that a transform is an analytic function of its independent variable, while Titchmarsh (1939) discuses this somewhat more generally. In essence, the transform is an analytic function of its variable β for those regions of the complex β plane wherein the integral is uniformly convergent. Also note that the condition that $u_3 = 0$ for $x_1 < 0$, $x_2 = 0$ has been built into ${}^*u_{3+}$.

Working next with (5.91), we find that

$$\begin{aligned} {}^*\tau_+ &= \int_0^\infty \tau_+(x_1)e^{-i\beta x_1}\,dx_1 \\ &= \frac{-A}{(\beta - i\epsilon)}. \end{aligned} \tag{5.94}$$

Note the pole at $i\epsilon$. Further we define the transform ${}^*\tau_-$ as

$${}^*\tau_- = \int_{-\infty}^0 \tau_-(x_1)e^{-i\beta x_1}\,dx_1. \tag{5.95}$$

The crack tip is the only possible source for a scattered wave in $x_1 < 0$. A cylindrical scattered wave is emitted from the tip. Thus $\tau_- = O(e^{-\epsilon|x_1|})$ as $kx_1 \to -\infty$, because $k' = k + i\epsilon$, from which it follows that ${}^*\tau_-$ is an analytic function of β for $\Im(\beta) > -\epsilon$.

We demand that $\partial_2 u_3$ calculated from (5.92) be consistent with that represented by (5.94) and (5.95) at $x_2 = 0$. This gives

$$i\gamma\,{}^*u_{3+} = [-A/(\beta - i\epsilon)] +{}^* \tau_-, \tag{5.96}$$

which is the functional relation we are looking for. Note that different parts of it are analytic in different parts of the complex β plane, but that all parts are analytic in the common strip $\Im(\beta) \in (-\epsilon, \epsilon)$. The goal of the next few paragraphs will be to rearrange this expression so that one side is analytic in the region $\Im(\beta) > -\epsilon$, while the other is analytic in the region $\Im(\beta) < \epsilon$. The two sides will

[6] We need to guess at the kinematics of the scattered wave to deduce where ${}^*u_{3+}$ and ${}^*\tau_-$ are analytic. Experience with similar problems is how this is done. However, the final answer must be checked to ensure that it is consistent with these assumptions.

be equal in the common strip and hence will be different representations of the same function, which function must then be entire. This process is the *Wiener–Hopf method.* Proceeding to this relation directly from the transforms, rather than through first forming an integral equation, is the simplification introduced by Jones (1952).

We next write (5.96) as

$$i(k' - \beta)^{1/2}\, {}^*u_{3+} = \frac{-A}{(\beta - i\epsilon)(k' + \beta)^{1/2}} + \frac{{}^*\tau_-}{(k' + \beta)^{1/2}}, \tag{5.97}$$

so that its left side becomes an analytic function for $\Im(\beta) < \epsilon$. The right side is almost one for $\Im(\beta) > -\epsilon$, but for the presence of the pole at $i\epsilon$. To isolate the pole term we add and subtract the residue multiplied by $(\beta - i\epsilon)^{-1}$. The outcome is

$$i(k' - \beta)^{1/2}\, {}^*u_{3+} + \frac{A}{(\beta - i\epsilon)(k' + i\epsilon)^{1/2}}$$

$$= -A\left[\frac{(k' + i\epsilon)^{1/2} - (k' + \beta)^{1/2}}{(\beta - i\epsilon)(k' + \beta)^{1/2}(k' + i\epsilon)^{1/2}}\right] + \frac{{}^*\tau_-}{(k' + \beta)^{1/2}}. \tag{5.98}$$

We have succeeded in achieving our goal. The left side is analytic for $\Im(\beta) < \epsilon$ and the right side for $\Im(\beta) > -\epsilon$, while the two sides are equal in the common strip $\Im(\beta) \in (-\epsilon, \epsilon)$. The two sides are thus the analytic continuations of one another, and together they represent a function analytic everywhere in the finite β plane. The function is entire and its nature is determined by its behavior at infinity.

We have still not used the edge condition. This is the condition that tells us how the entire function behaves at infinity. Recall that we asked that $ku_{3+} = \mathrm{O}[(k|x_1|)^{1/2}]$ as $k|x_1| \to 0$. From this we may infer that $\tau_- = \mathrm{O}[(k|x_1|)^{-1/2}]$ as $k|x_1| \to 0$. Adapting an Abelian theorem $k^2\, {}^*u_{3+} = \mathrm{O}(k^{3/2}|\beta|^{-3/2})$ and that $k\, {}^*\tau_- = \mathrm{O}(k^{1/2}|\beta|^{-1/2})$ as $k^{-1}|\beta| \to \infty$. Using Liouville's theorem (Titchmarsh, 1939) it follows that the entire function must be identically zero. Therefore

$${}^*u_{3+} = \frac{iA}{(\beta - i\epsilon)(k' + i\epsilon)^{1/2}(k' - \beta)^{1/2}}. \tag{5.99}$$

Note how the edge condition was essential to the determination of a unique solution.

At this point ϵ has served its purpose and we take the limit $\epsilon \to 0$. We are thus left with an integral expression for the scattered wavefield, namely

$$u_3 = \frac{iA}{2\pi k^{1/2}} \int_{-\infty}^{\infty} \frac{e^{i(\beta x_1 + \gamma x_2)}}{\beta (k-\beta)^{1/2}}\, d\beta, \tag{5.100}$$

where the contour passes below $\beta = 0$. Integrals of this form or of the form achieved after using the Sommerfeld transformation are sometimes referred to as diffraction integrals. In closing, we recall that this is the solution to the scattered wavefield only for $x_2 > 0$.

Problem 5.5 An Arbitrary Angle of Incidence

Problem 1. Repeat the calculations leading to (5.100) for an arbitrary angle of incidence θ_0. The incident wave is given by

$$u_3^i = (A/k) e^{ik'(\cos\theta_0 x_1 + \sin\theta_0 x_2)}, \tag{5.101}$$

where $k' = k + i\epsilon$ and $\epsilon > 0$. Initially assume that $\theta_0 < \pi/2$. Why might this be a useful restriction? Once the solution is obtained, can the restriction be removed? Show that the scattered wavefield u_3 is given by

$$u_3 = \frac{iA}{2\pi k^{1/2}} \frac{\sin\theta_0 e^{ik(\cos\theta_0 x_1)}}{(1+\cos\theta_0)^{1/2}} \int_{-\infty}^{\infty} \frac{e^{i(\beta x_1 + \gamma x_2)}}{(\beta - k\cos\theta_0)(k-\beta)^{1/2}}\, d\beta. \tag{5.102}$$

Problem 2. The integral (5.102) has a pole at $\beta = k\cos\theta_0$. Calculate an asymptotic approximation to (5.102) that is not uniform. Include the pole term when it it is needed and note when the asymptotic approximation breaks down.

When we calculate a steepest descents approximation to (5.102), we must take care that the stationary point does not lie near the pole, because in that case we can no longer argue that the integrand, aside from the exponential term, is slowly varying. An asymptotic approximation, calculated when one parameter is large, that becomes disordered for certain values of a second parameter, in this case the angle of incidence θ_0, is said *not* to be *uniform*. It is uniform when the approximation is accurate for all values of the second parameter.

5.5.3 Description of the Scattered Wavefield

We have indicated five regions in Fig. 5.6, each of which has a somewhat different wavefield. In region 1 the wavefield u_3 is composed of a residue term, from the pole in the integrand of (5.100) at $\beta = 0$, plus the integral itself. The residue contribution cancels the u_3^i so that the crack casts the region behind it into partial silence. The integral represents a diffracted wave that appears as a cylindrical one radiating from the crack tip. This is the only wave that penetrates the silence. In region 2 the wavefield u_3 consists solely of the integral representing the diffracted wave, though the total wavefield would include u_3^i as well. In region 3 the wavefield u_3 is again composed of a residue term plus the integral itself. However, note that in using the antisymmetry, namely that $u_3(x_1, x_2) = -u_3(x_1, -x_2)$ to calculate u_3 for $x_2 < 0$, the residue contribution gives the wave reflected from the crack, while the integral continues to represent the diffracted wave. The incident wave u_3^i must be added to this to produce the total wavefield. Regions 4 and 5 are transition or boundary layers. Moving from left to right through these layers, the scattered wavefield makes a transition from solely a diffracted wave to a diffracted plus geometrical wave. These regions are characterized by Fresnel integrals, as we subsequently demonstrate. Within these regions the diffracted wave comes close to phase matching to the geometrical wave so that the two kinds of waves strongly interact, thereby producing a complicated wavefield, but propagate independently elsewhere.

Note that (5.100) contains both the diffracted wave and geometrical terms. A uniform asymptotic approximation to (5.100) is calculated in Felsen and Marcuvitz (1994).[7] However, most of the manipulations undertaken to do so are no more difficult than those needed to reduce (5.100) to an exact combination of Fresnel integrals. This reduction is described in detail in the Appendix. The Fresnel integral is defined as

$$F(z) := \int_z^\infty e^{i\xi^2}\, d\xi. \tag{5.103}$$

We introduce the polar coordinates (r, θ), where $x_1 = r\cos\theta$ and $x_2 = r\sin\theta$. For $x_2 > 0$, $\theta \in (0, \pi]$, while for $x_2 < 0$, $\theta \in (0, -\pi]$. The outcome of the calculations described in the Appendix is that

$$u_3 = \frac{e^{-i\pi/4} A}{\pi^{1/2} k}\left\{ e^{-ikr\cos(\theta-\pi/2)} F\left[(2kr)^{1/2}\cos\left(\frac{\theta-\pi/2}{2}\right)\right] \right.$$
$$\left. - e^{-ikr\cos(\theta+\pi/2)} F\left[-(2kr)^{1/2}\cos\left(\frac{\theta+\pi/2}{2}\right)\right]\right\}. \tag{5.104}$$

It is important to note that this expression assumes that $x_2 > 0$ or, equivalently,

[7] The calculation is done by separating a term that contains the pole from the rest of the integrand. One way to do this is to subtract and add the integrand of (5.125). The integral with no pole term in its integrand is approximated in the usual way, while the integral with the pole term becomes a Fresnel integral.

$\theta \in (0, \pi]$. To obtain an expression for $x_2 < 0$, we use the fact that $u_3(r, \theta) = -u_3(r, -\theta)$.

To calculate the total wavefield, we use the following property of the Fresnel integral:

$$F(z) + F(-z) = \pi^{1/2} e^{i\pi/4}. \tag{5.105}$$

The total wavefield u_3^t, for $\theta \in (-\pi, \pi]$, is, after adding u_3^i to (5.104), given by

$$u_3^t = \frac{e^{-i\pi/4} A}{\pi^{1/2} k} \left\{ e^{-ikr\cos(\theta-\pi/2)} F\left[(2kr)^{1/2} \cos\left(\frac{\theta - \pi/2}{2} \right) \right] \right.$$
$$\left. + e^{-ikr\cos(\theta+\pi/2)} F\left[(2kr)^{1/2} \cos\left(\frac{\theta + \pi/2}{2} \right) \right] \right\}. \tag{5.106}$$

For $kr \gg 1$ these integrals can be asymptotically approximated. *Problem 5.7* describes the asymptotic approximations to the Fresnel integral. When this is done we find, for $\theta \in (0, \pi]$, that

$$u_3^t \sim \frac{A}{k} e^{ikx_2} H(-x_1) + D(\theta, \pi/2) \frac{A}{k} \frac{e^{ikr}}{(kr)^{1/2}}, \qquad kr \to \infty, \tag{5.107}$$

where

$$D(\theta, \pi/2) = \frac{e^{i\pi/4}}{2(2\pi)^{1/2}} \left\{ \frac{1}{\cos[(\theta - \pi/2)/2]} + \frac{1}{\cos[(\theta + \pi/2)/2]} \right\}. \tag{5.108}$$

We find a very similar expression for $\theta \in (0, -\pi]$. Note the similarity in structure between the expansion (5.107) and the earlier expansion (2.41). The coefficient $D(\theta, \pi/2)$ is called a diffraction coefficient. Its arguments are intended to suggest that a ray incident to the crack tip at an angle $\pi/2$ excites a diffracted ray that leaves the tip at an angle θ. In fact, there are a whole fan of such rays as θ is allowed to take on all its values. These coefficients play an important role in the geometrical theory of diffraction for edges. The interested reader can pursue this theory further in books by Achenbach et al. (1982), Babič and Buldyrev (1991), and Jull (1981). *Problem 5.6* introduces some of the ideas used in this description of diffraction.

The integrals of (5.106) can also be approximated for $kr \ll 1$, as *Problem 5.7* indicates. Any singular behavior as $kr \to 0$ is contained entirely within u_3, whose approximation in this limit is

$$ku_3 = \mathrm{O}[(kr)^{1/2}], \qquad kr \to 0. \tag{5.109}$$

This reproduces the edge condition we enforced in (5.89).

The approximation (5.107) breaks down near $\theta = \pm\pi/2$. Examining (5.106), we note that one or the other of the arguments of the Fresnel functions changes sign at these angles. Let us consider the neighborhood of $\pi/2$ so that it is the second Fresnel integral in (5.106) whose argument changes sign. Examining this argument, we find that if θ is close to $\pi/2$ the argument of the Fresnel integral is too small to approximate it asymptotically, despite kr being quite large. Thus if

$$(2kr)^{1/2} \cos\left(\frac{\theta + \pi/2}{2}\right) \leq 1, \tag{5.110}$$

such an approximation is not accurate. Taking the equality as forming the boundary of the region outside of which the asymptotic approximation is sufficiently accurate, we find that the boundary is described by $kr(1 - \sin\theta) = 1$, the equation of a parabola in polar form. This then marks the boundary of region 4 with regions 1 and 2. The boundary layer lies within this boundary, and its scale is set by the expression on the left in (5.110). The boundary of region 5 is simply the reflection in $x_2 = 0$ of that for region 4.

Problem 5.6 The Geometrical Theory of Diffraction

Consider an antiplane shear wave normally incident to the opening between two cracks, as shown in Fig. 5.9. Estimate the wavefield diffracted by the strip by using what has been learned from the semi-infinite crack problem. How does this problem, aside from the fact that the time dependence is harmonic, differ from *Problem 5.1*?

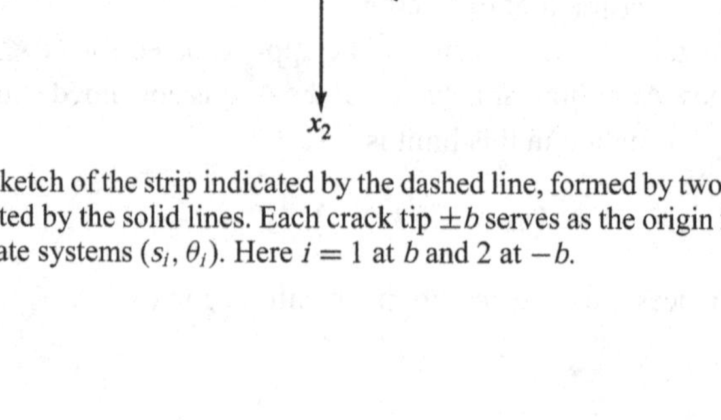

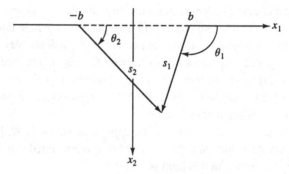

Fig. 5.9. A sketch of the strip indicated by the dashed line, formed by two semi-infinite cracks, indicated by the solid lines. Each crack tip $\pm b$ serves as the origin for one of the polar coordinate systems (s_i, θ_i). Here $i = 1$ at b and 2 at $-b$.

As before, express the total wavefield as $u_3^t = u_3^i + u_3$. The incident wave is given by (5.84). Are the boundary and continuity conditions at $x_2 = 0$ the following?

$$\mu\partial_2 u_3^t(x_1, 0^{\pm}) = 0, \quad x_1 \in (-\infty, -b) \cup (b, \infty),$$
$$u_3^t(x_1, 0^+) = u_3^t(x_1, 0^-), \quad x_1 \in (-b, b). \tag{5.111}$$

Formulate the problem for the scattered wavefield u_3, in $x_2 > 0$, stating all the conditions that must be satisfied. An exact analytic solution to this problem may not be possible. Explain, using what has been learned from the semi-infinite crack problem, why the following expression is a plausible asymptotic solution to this problem

$$u_3^t = \frac{A}{k}[H(x_1 + b) - H(x_1 - b)]e^{ikx_2}$$
$$+ D\left(\pi - \theta_2, \frac{\pi}{2}\right)\frac{A}{k}\frac{e^{iks_2}}{(ks_2)^{1/2}} + D\left(\theta_1, \frac{\pi}{2}\right)\frac{A}{k}\frac{e^{iks_1}}{(ks_1)^{1/2}}, \tag{5.112}$$

where the coordinates (s_1, θ_1) and (s_2, θ_2) are defined in Fig. 5.9 and $D(\theta, \pi/2)$ is given by (5.108). This solution is best for $2kb \gg 1$. Why?

Consider the boundary layers surrounding $\theta_{1,2} = \pi/2$. They grow in width as one moves away from the edges. At what point do they choke off the geometrical term given by the first term of (5.112)? The earlier parts of Harris (1987) might be useful, but are not in any way essential, in answering this last question.

5.6 Matched Asymptotic Expansion Study

We have just undertaken a relatively complex solution to what, looking at Fig. 5.7, might seem a simple wave process. After all, to some leading order of approximation, the crack merely blocks the incident wave from reaching the other side. In this closing section we construct an asymptotic solution that captures this idea mathematically. In writing this I have followed a similar discussion in Zauderer (1983) and benefited from a very detailed analysis of edge diffraction by Gautesen (1979).

In Section 2.4 we explored how waves could be approximated by a ray theory and provided an asymptotic structure that explained the connection. Using this structure, we begin our analysis of the diffraction problem by assuming that the wavefield can be described with the asymptotic approximation

$$u_3 \sim e^{ikS(x_1,x_2)}\sum_{n\geq 0}(-ik)^{-n}A_n(x_1, x_2). \tag{5.113}$$

This assumption leads to the eikonal equation (2.43) for S and the transport equation (2.44) for A_0.

First, we consider the region $x_1 > 0$. To match the phase of the incident wave u_3^i at $x_2 = 0$, given by (5.84), we ask that $S(x_1, 0) = 0$ and that the scattered waves be outgoing from the crack. The appropriate solution is $S(x_1, x_2) = \pm x_2\, H(x_1)$, where the plus sign is used for $x_2 > 0$ and the minus sign for $x_2 < 0$, and $H(x_1)$ is the Heaviside function. Knowing S, the transport equation, (2.44), indicates that A_0 can have at most an x_1 dependence. Moreover, it must be such that the traction for the total wavefield vanishes on both sides of the crack. The appropriate solution is $A_0(x_1) = \mp(A/k)\, H(x_1)$, where the minus sign is used for $x_2 > 0$ and the plus sign for $x_2 < 0$. It also follows that $A_n \equiv 0$ for $n \geq 1$. Second, we consider the region $x_1 < 0$. There are no boundary conditions to enforce, leading to the conclusion that $A_n \equiv 0$ for $n \geq 0$. Therefore, to leading order the total wavefield u_3^t is given by

$$u_3^t(x_1, x_2) = \begin{cases} (A/k)e^{ikx_2}, & \text{region 2}, \\ (A/k)(e^{ikx_2} + e^{-ikx_2}), & \text{region 3}, \\ 0, & \text{region 1}, \end{cases} \tag{5.114}$$

where the regions are those indicated in Fig. 5.7.

This solution gives the geometrical part of (5.106), but does not capture the transition in regions 4 and 5. Even in the absence of an exact solution, we might suspect that our asymptotic ansatz does not have enough structure to capture the effects of the rapid changes in the wavefield near $x_1 = 0$ just from the apparent discontinuity there. We have encountered a boundary layer. What we need to do is scale the problem in such a way that the boundary layer is opened up and an equation governing the wavefield within it found. The scaling needed to open the layer up is usually not known and must be found as part of discovering the governing equation. We then construct the solution to this equation so that it matches the surrounding wavefield to some order of approximation. Such a procedure is called *matched asymptotic expansions.* Both Hinch (1991) and Holmes (1995) discuss this technique thoroughly.

Moreover, recall that, from our analysis in Section 4.7.3, we know that very near the crack tip the wavefield is quasi-static, with part of it behaving as $(kr)^{1/2}$. In addition to the boundary layer, that we are about to examine, there is a nearfield or inner region that is not described by the asymptotic anzatz (5.113). However, for the present we avoid this nearfield.

We consider the case that $x_2 > 0$ so that θ is near $\pi/2$ and leave the case $x_2 < 0$ to the reader. Examining (5.106), we note that in this region a propagator term e^{ikx_2} always appears. Accordingly, we set

$$u_3 = w(x_1, x_2)e^{ikx_2} \tag{5.115}$$

and arrive at the following equation for w:

$$2ik\partial_2 w + \partial_2^2 w + \partial_1^2 w = 0. \tag{5.116}$$

We have stripped away the oscillatory part of the wavefield. At this point we note that if w varies rapidly in the neighborhood of $x_1 = 0$, then $\partial_1^2 w$ may be very large and to solve the equation we shall need another term to balance it. The term $2ik\partial_2 w$ seems a likely candidate because it is multiplied by the large parameter k.

We have indicated previously that a length scale is very important, as it sets a gauge to measure large and small. The diffraction problem does not have a natural length scale, so that we must introduce one somewhat artificially, just as we did when examining the local field near the crack tip in Section 4.7.3. We again take the length ϵ as a reference length. It might represent a wavelength at a reference frequency or a distance at which a measurement is made. We then define the nearfield as that for which $k\epsilon \ll 1$ and the farfield as that for which $k\epsilon \gg 1$. We are examining the wavefield in the region $k\epsilon \gg 1$ with θ near $\pi/2$.

We scale the problem by setting $\bar{x}_i = x_i/\epsilon$ and $\bar{w} = wk/A$ (recall that A/k is a magnitude of the incident wave) and reexpress (5.116) in terms of the scaled variables. Having done this, we now *omit the overbar.* Equation (5.116) has become

$$2i(k\epsilon)\partial_2 w + \partial_2^2 w + \partial_1^2 w = 0. \tag{5.117}$$

The argument, expressed in terms of the scaled (x_1, x_2), of the second Fresnel function in (5.106), namely $-x_1(k\epsilon/2x_2)^{1/2}$, suggests the scaling needed to open up the boundary layer. We set $y_1 = (k\epsilon)^\beta x_1$ with $\beta = 1/2$. This change of coordinate is called *introducing a stretching transformation.* Note, however, that we do not in general know β and must determine it from the rescaled equation by balancing the various terms (Hinch, 1991; Holmes, 1995). In this case it is the first and third terms in (5.117) that balance. The coordinates (y_1, x_2) are called the inner coordinates and the (scaled) (x_1, x_2) the outer coordinates.

We next introduce an asymptotic expansion of the form

$$w \sim w_0(y_1, x_2) + (k\epsilon)^{\gamma} w_1(y_1, x_2) \ldots, \tag{5.118}$$

where $\gamma > 0$. The leading-order term is governed by the parabolic Schrodinger equation,

$$2i(\partial w_0/\partial x_2) + (\partial^2 w_0/\partial y_1^2) = 0. \tag{5.119}$$

This is the equation governing the behavior of the wavefield in the boundary layer.

Recalling the fundamental or causal Green's function for the diffusion equation (Zauderer, 1983), we write the solution to (5.119) as

$$w_0 = \frac{1}{x_2^{1/2}} \int_{-\infty}^{\infty} f\big[(k\epsilon)^{1/2} s\big] e^{i(y_1 - s)^2/(2x_2)}\, ds. \tag{5.120}$$

To find the unknown $f(x)$ we express this equation in terms of the outer variables as

$$w_0 = \frac{(k\epsilon)^{1/2}}{x_2^{1/2}} \int_{-\infty}^{\infty} f(s) e^{ik\epsilon(x_1 - s)^2/(2x_2)}\, ds. \tag{5.121}$$

The reader should recall that we are assuming $x_2 > 0$. This integral can be asymptotically expanded for large $k\epsilon$, provided $f(s)$ does not vary rapidly near the stationary point $s = x_1$. However, the function $f(s)$ must vary rapidly near $x_1 = 0$ if it is to capture the transition in the boundary layer, so that an asymptotic approximation there will not be accurate. However, for $|x_1| > 0$ we expect that the asymptotic approximation to (5.121) should match our earlier approximation to u_3. That is, $w_0 \sim 0$ when $x_1 < 0$ and $w_0 \sim -1$ for $x_1 > 0$. Using the stationary phase approximation, (5.69), or the steepest descents approximation, we find that

$$w_0 \sim f(x_1)(2\pi)^{1/2} e^{i\pi/4}. \tag{5.122}$$

Thus we find for $x_2 > 0$ that

$$f(x_1) = -e^{-i\pi/4}/(2\pi)^{1/2} H(x_1). \tag{5.123}$$

This process of finding $f(x_1)$ is referred to as *matching the inner and outer expansions.* At this point we could now match the nearfield expansion (4.52) to (5.120) to determine unequivocally that $\beta = 1/2$ and also to determine the unknown constant A in the nearfield expansion (4.52) (Gautesen, 1979).

Removing all the scaling so that we may compare our result with (5.104), we find that

$$u_3 \sim -\frac{Ae^{-i\pi/4}}{k\pi^{1/2}} e^{-ikx_2} F\left[-x_1\left(\frac{k}{2x_2}\right)^{1/2}\right], \tag{5.124}$$

where $F(z)$ is the Fresnel integral defined previously by (5.103). For $x_2 > 0$ and $x_1 \approx 0$, (5.104) reduces to this expression. If we need a boundary layer solution for $x_2 < 0$, region 5, we use the antisymmetry of the scattered wavefield, namely $u_3(x_1, x_2) = -u_3(x_1, -x_2)$, which is a global property of the solution.

Had we expanded (5.121) to a second term, we should have encountered a term $O[(k\epsilon)^{-1/2}]$ but found no similar term in the expansion (5.113) to match it. Why? The ansatz (5.113) contains only the geometrical wavefield and not the diffracted one. Examining (5.107), we see immediately that the missing term comes from the diffracted wavefield. To include this possibility we should have allowed for fractional powers of k in positing (5.113) just as we did with the α in (2.41). Had we not known the solution nor suspected what was going on, the failure to match at higher order would have indicated that something was missing.

Appendix: The Fresnel Integral

Our purpose is to derive the expression (5.104). The integral to be evaluated is (5.100), which we rewrite here.

$$u_3 = \frac{iA}{2\pi k^{1/2}} \int_{C_\beta} \frac{e^{i(\beta x_1 + \gamma x_2)}}{\beta(k-\beta)^{1/2}}\, d\beta. \tag{5.125}$$

An integral that has a pole and stationary point (usually a saddle), points that may coalesce for certain values of the physical coordinates, is called a diffraction integral. The essence of its evaluation is to reduce it to one on its contour of steepest descents. I follow the derivation given in Born and Wolf (1986). It is repeated here so that the calculation begun in Section 5.5 can be completed.

Recall that the diffraction integral is evaluated along the contour C_β that passes below the pole at $\beta = 0$. We use the Sommerfeld transformation $\beta = k\cos\alpha$ and $\gamma = k\sin\alpha$ and introduce the coordinates (r, θ) by setting $x_1 = r\cos\theta$ and $x_2 = r\sin\theta$ to reduce (5.125) to an integral over the contour C in the α plane, namely

$$u_3 = -\frac{iA}{2^{1/2}\pi k} \int_C \frac{\cos(\alpha/2)}{\cos\alpha} e^{ikr\cos(\alpha-\theta)}\, d\alpha. \tag{5.126}$$

Note that the contour now passes above the pole at $\alpha = \pi/2$.

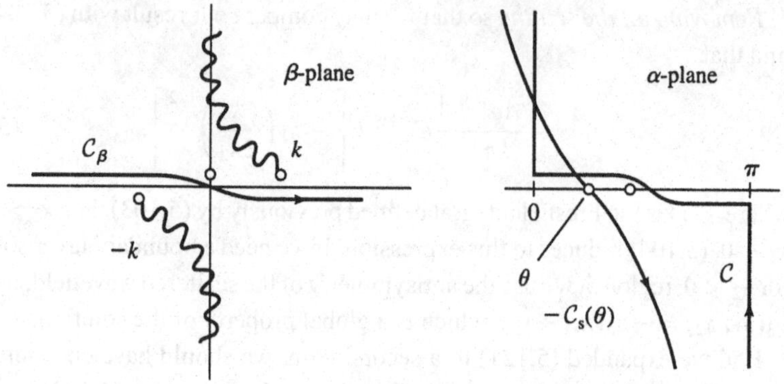

Fig. 5.10. The complex β plane for the diffraction integral, showing the contour C_β, is sketched on the left. The complex α plane, showing the contours C and $-C_s(\theta)$, is sketched on the right. θ is the saddle point in the α plane.

The following two identities are needed for the work to follow:

$$\frac{2^{3/2}\cos(\alpha/2)}{\cos\alpha} \equiv \frac{1}{\cos[(\alpha-\pi/2)/2]} + \frac{1}{\cos[(\alpha+\pi/2)/2]}, \tag{5.127}$$

$$\frac{4\cos(\alpha/2)\cos[(\theta-\pi/2)/2]}{\cos\alpha+\cos(\theta-\pi/2)} \equiv \frac{1}{\cos[(\alpha+\theta-\pi/2)/2]} + \frac{1}{\cos[(\alpha-\theta+\pi/2)/2]}. \tag{5.128}$$

The second identity is a generalization of the first.

We now distort the contour C to the steepest descents contour C_s described by (5.76) and (5.77). We next reverse the direction of integration, writing the symbol for the contour as $-C_s(\theta)$. This contour is shown in Fig. 5.10. The position of the stationary point is indicated as an argument of this contour to underscore the structure of the contour. The diffraction integral is now written as

$$u_3 = \frac{iA}{4\pi k}\int_{-C_s(\theta)} e^{ikr\cos(\alpha-\theta)} \times \left\{\frac{1}{\cos[(\alpha-\pi/2)/2]} + \frac{1}{\cos[(\alpha+\pi/2)/2]}\right\} d\alpha, \tag{5.129}$$

where (5.127) has been used.

Setting aside the $iA/(4\pi k)$, we consider the first of the two integrals, which we label I. We note that the calculation of the second is not different from that of the first. We next change variables, giving

$$I = \frac{1}{2}\int_{-\mathcal{C}_s(0)} e^{ikr\cos\alpha}$$

$$\times \left\{ \frac{1}{\cos[(\alpha+\theta-\pi/2)/2]} + \frac{1}{\cos[(\alpha-\theta+\pi/2)/2]} \right\} d\alpha. \quad (5.130)$$

Using (5.128), we can collapse this integral into the more symmetric form

$$I = 2\int_{-\mathcal{C}_s(0)} \frac{\cos(\alpha/2)\cos[(\theta-\pi/2)/2]}{\cos\alpha + \cos(\theta-\pi/2)} e^{ikr\cos\alpha}\, d\alpha. \quad (5.131)$$

We now introduce yet another change of variables with the relation $\tau = 2^{1/2} e^{i\pi/4}\sin(\alpha/2)$ (recall *Problem 5.3*). Thus the integral I becomes

$$I = -2e^{i\pi/4} e^{ikr}\eta \int_{-\infty}^{\infty} \frac{e^{-kr\tau^2}}{\tau^2 - i\eta^2}\, d\tau, \quad (5.132)$$

where $\eta = 2^{1/2}\cos[(\theta-\pi/2)/2]$.

A third identity, namely

$$\int_{kr}^{\infty} e^{i\eta^2\zeta} \int_{-\infty}^{\infty} e^{-\zeta\tau^2}\, d\tau\, d\zeta \equiv \pi^{1/2} \int_{kr}^{\infty} \zeta^{-1/2} e^{i\eta^2\zeta}\, d\zeta, \quad (5.133)$$

is needed. Using this (interchange the order of integration on the left-hand side), we find that

$$\eta \int_{-\infty}^{\infty} \frac{e^{-kr\tau^2}}{\tau^2 - i\eta^2} d\tau = \frac{2\pi^{1/2}\eta}{|\eta|} e^{-ikr\eta^2} \int_{|\eta|(kr)^{1/2}}^{\infty} e^{i\mu^2}\, d\mu'$$

$$= \frac{2\pi^{1/2}\eta}{|\eta|} e^{-ikr\eta^2} F\left[|\eta|(kr)^{1/2}\right], \quad (5.134)$$

where the Fresnel function $F(z)$, repeating the definition (5.103), is

$$F(z) := \int_{z}^{\infty} e^{i\xi^2}\, d\xi. \quad (5.135)$$

We now return to (5.129). Note that this integral contains both a pole contribution and a part composed of Fresnel integrals. Let u_{3np} represent the particle displacement that does not include the pole contribution and u_{3p} the pole contribution itself. Now u_{3np} is given by

$$u_{3np} = \frac{e^{-i\pi/4} A}{\pi^{(1/2)} k} \left\{ e^{-ikr\cos(\theta-\pi/2)} F\left[(2kr)^{1/2}\cos\left(\frac{\theta-\pi/2}{2}\right)\right]\right.$$

$$\left. \pm\, e^{-ikr\cos(\theta+\pi/2)} F\left[\pm(2kr)^{1/2}\cos\left(\frac{\theta+\pi/2}{2}\right)\right]\right\}, \quad (5.136)$$

where the plus sign is used for $\theta \in (0, \pi/2)$ and the minus sign for $\theta \in (\pi/2, \pi)$. Here u_{3p} is given by

$$u_{3p} = -(A/k)e^{-ikr\cos(\theta-\pi/2)} H(\pi/2 - \theta). \tag{5.137}$$

Recalling (5.105),

$$F(z) + F(-z) = \pi^{1/2} e^{i\pi/4}, \tag{5.138}$$

we use it to combine (5.137) with the second term of (5.136) to give (5.104).

Problem 5.7 Asymptotic Approximations to the Fresnel Integral

Consider the Fresnel integral (5.135) for $z = x$, where x is real.

Problem 1. Find the first term of an asymptotic expansion of the Fresnel integral for both positive and negative x.

(a) Assume $x > 0$ and consider the integral

$$G(x) = e^{-ix^2} \int_x^\infty e^{i\zeta^2}\, d\zeta. \tag{5.139}$$

Show that

$$G(x) \sim \frac{i}{2x} + O(x^{-3}), \qquad x \to \infty. \tag{5.140}$$

One way to do this is to deform the contour to one such that *Watson's lemma* can be used.

(b) Assume $x < 0$. The approach of part (a) must be modified. Why? Write the integral in (5.139) as one from x to $-\infty$ and one from $-\infty$ to ∞. Do not use (5.138). Hence show that

$$G(x) \sim \pi^{1/2} e^{i\pi/4} e^{-ix^2} + \frac{i}{2x} + O(x^{-3}), \qquad x \to -\infty. \tag{5.141}$$

That the two approximations differ for x that is positive or negative is an example of the *Stokes' phenomena.*

Problem 2. Find the first term of an asymptotic expansion of the Fresnel integral for $|x| \ll 1$. Write the integral as the difference between one from zero to ∞ and one from zero to x. Expand the integrand of the second integral in a Taylor expansion. Hence show that

$$F(x) \sim (\pi^{1/2} e^{i\pi/4})/2 - x + O(x^3), \qquad x \to 0. \tag{5.142}$$

References

Ablowitz, M.J. and Fokas, A.S. 1997. *Complex Variables*, pp. 411–513. New York: Cambridge.

Achenbach, J.D., Gautesen, A.K., and McMaken, H. 1982. *Ray Methods for Waves in Elastic Solids*. Boston: Pitman.

Babič, V.M. and Buldyrev, V.S. 1991. *Short-Wavelength Diffraction Theory*. Berlin: Springer.

Born, M. and Wolf, E. 1986. *Principles of Optics*, 6th (corrected) ed., pp. 565–578. Oxford: Pergamon.

Cagniard, L. 1962. *Reflection and Refraction of Progressive Seismic Waves*. Translated and revised by E.A. Flinn and C.H. Dix. New York: McGraw-Hill.

Carrier, G.F., Krook, M., and Pearson, C.E. 1983. *Functions of a Complex Variable*, pp. 249–283. Ithaca, NY: Hod Books.

Copson, E.T. 1935. *An Introduction to the Theory of Functions of a Complex Variable*, pp. 121–125. Oxford: Clarendon Press.

Copson, E.T. 1971. *Asymptotic Expansions*, pp. 13–14 and 48–62. Cambridge: University Press.

Courant, R. and John, F. 1989. *Introduction to Calculus and Analysis*, Vol. II, pp. 345–350. New York: Springer

deHoop, A.T. 1960. A modification of Cagniard's method for solving the seismic pulse problem. *Appl. Sc. Res.*, B **8**: 349–356.

Doetsch, G. 1974. *Introduction to the Theory and Application of the Laplace Transform*, pp. 218–230. New York: Springer.

Ewing, W.M., Jardetzky, W.S., and Press, F. 1957. *Elastic Waves in Layered Media*. New York: McGraw-Hill.

Felsen, L.B. and Marcuvitz, N. 1994. *Radiation and Scattering of Waves*, pp. 370–441. New York: IEEE and Oxford University Presses.

Gautesen, A.K. 1979. On matched asymptotic expansions for two dimensional elastodynamic diffraction by cracks. *Wave Motion* **1**: 127–140.

Harris, J.G. 1980a. Diffraction by a crack of a cylindrical longitudinal pulse. *Z. Angew. Math. Phys.* **31**: 367–383. Errata, *Z. Angew. Math. Phys.* **34**.

Harris, J.G. 1980b. Uniform approximations to pulses diffracted by a crack. *Z. Angew. Math. Phys.* **31**: 771–775.

Harris, J.G. 1987. Edge diffraction of a compressional beam. *J. Acoust. Soc. Am.* **82**: 635–646.

Harris, J.G. and Pott, J. 1985. Further studies of scattering of a Gaussian beam from a fluid-solid interface. *J. Acoust. Soc. Am.* **78**: 1072–1080.

Hinch, E.J. 1991. *Perturbation Methods*, pp. 52–101. Cambridge: University Press.

Holmes, M.H. 1995. *Introduction to Perturbation Methods*, pp. 47–104. New York: Springer.

Jones, D.S. 1952. A simplifying technique in the solution of a class of diffraction problems. *Quart. J. Math.* **3**: 189–196.

Jull, E.V. 1981. *Aperture Antennas and Diffraction Theory*. London: Peter Peregrinus.

Knopoff, L. and Gilbert, F. 1959. First motion methods in theoretical seismology. *J. Acoust. Soc. Am.* **31**: 1161–1168.

Noble, B. 1988. *Methods Based on the Wiener-Hopf Technique*, pp. 11–27 and 48–97. New York: Chelsea.

Titchmarsh, E.C. 1939. *The Theory of Functions*, 2nd ed., pp. 85–86 and 99–101. Oxford: Clarendon Press.

Titchmarsh, E.C. 1948. *Introduction to the Theory of Fourier Integrals*, 2nd ed. Oxford: Clarendon Press.

van der Pol, B. and Bremmer, H. 1950. *Operational Calculus*, pp. 122–132 and elsewhere. Cambridge: University Press.

Wong, R. 1989. *Asymptotic Approximations of Integrals*, pp. 20–31 and elsewhere. Boston: Academic.

Zauderer, E. 1983. *Partial Differential Equations of Applied Mathematics*, pp. 653–658 and 403–405. New York: Wiley-Interscience.

6

Guided Waves and Dispersion

Synopsis

Chapter 6 discusses guided waves and the dispersion they experience. Only the antiplane shear problem is treated. The guided waves are constructed by using partial waves and their dispersion calculated by using the transverse resonance principle. Both harmonic and transient excitations of a closed waveguide are studied by using an expansion of modes. The harmonic excitation of an open waveguide by a line source is also studied, though in this case by using both ray and mode representations. As a last example, we examine propagation in a closed waveguide with a slowly varying thickness, using an asymptotic expansion that combines features of both rays and modes. We close by examining how information and energy propagate at the group velocity.

6.1 Harmonic Waves in a Closed Waveguide

We consider a layer of infinite extent in the x_1 direction and of finite thickness in the x_2 direction. Within the layer, the coordinate $x_2 \in (-h, h)$ and the plane $x_2 = 0$ is a plane of reflection symmetry. This structure is a waveguide or guide because the waves are forced to propagate in the x_1 direction and the guide is closed because waves are completely trapped within the structure. We are interested in learning what kinds of antiplane waves propagate in the guide without, at present, seeking to know how they are excited. Accordingly, we seek possible solutions to the following antiplane problem. In the layer, u_3 must satisfy (2.20), which, rewritten here, is

$$\partial_a \partial_a u_3 + k^2 u_3 = 0, \tag{6.1}$$

and at the surfaces

$$\mu \partial_2 u_3(x_1, \pm h) = 0. \tag{6.2}$$

As in previous chapters the subscript T has been dropped. *Problem 6.1* indicates how solutions to this problem can be found by reducing it to an eigenvalue problem in the transverse coordinate. When this is done we find that the possible wavefields are described by

$$u_{3m} = A_m \begin{matrix}\cos\\ \sin\end{matrix}(\gamma_m x_2) e^{i\beta_m x_1}, \qquad m = 0, 1, 2, \ldots, \tag{6.3}$$

where u_{3m} is the mth (waveguide) mode for this guide. Cosine is used for m even and sine for m odd. This wave stands in the x_2 direction, but it propagates in the x_1 direction. Here β_m is the lateral wavenumber for the mth mode. The transverse wavenumbers for this mode are $\pm\gamma_m$. $\beta_m = \omega/c_m$, where c_m is the phase velocity of the mode. β_m is given by

$$\beta_m = [k^2 - (m\pi)^2/(2h)^2]^{1/2}, \tag{6.4}$$

where $\Re(\beta_m) \geq 0$ or $\Im(\beta_m) \geq 0$ for $x_1 > 0$. In this particular case β_m is either real or imaginary, but not complex. For β_m that is real, the wavenumber and phase velocity in the propagation direction depend on ω (through $k = \omega/c$). Thus (6.4) is a *dispersion relation* and is similar to that found previously for propagation in periodic structures (there we used κ to represent the effective wavenumber). The square root is defined so that, for imaginary values, the mode decays in its direction of propagation ($x_1 > 0$).

Introducing the term[1] $e^{i\beta x_1}$ will reduce a linear partial differential equation, or system of such equations, in (x_1, x_2) to an eigenvalue problem in x_2 for β, though that problem may not be easy to solve, or its solution may not be particularly informative. However, by examining the solution, (6.3), we can learn to reconstruct it in a way that uses only the kinematical features of reflection and transmission of a plane wave at a boundary. Thereafter this method of solution can be used to solve other guided wave problems without directly considering the underlying eigenvalue problem.

Problem 6.1 An Eigenvalue Problem: A Closed Waveguide

Equation (6.1), with boundary condition (6.2), can be reduced to an ordinary differential equation in x_2 by seeking solutions of the form

$$u_3 = f(x_2) e^{i\beta x_1}. \tag{6.5}$$

[1] This is equivalent to taking the spatial Fourier transform, transform variable β, in the x_1 direction.

Show that the differential equation in x_2 leads to an eigenvalue problem for β^2. Usually the β is linked with γ, where

$$\gamma = (k^2 - \beta^2)^{1/2}, \tag{6.6}$$

and γ^2 or $-\gamma^2$ is considered the eigenvalue. In this book β^2 is considered the eigenvalue. Show that solutions symmetric with respect to the reflection plane are

$$f_{2n}(x_2) = A_{2n} \cos \gamma_{2n} x_2, \qquad \gamma_{2n} = n\pi/h, \tag{6.7}$$

and that those antisymmetric to this plane are

$$f_{2n+1}(x_2) = A_{2n+1} \sin \gamma_{2n+1} x_2, \quad \gamma_{2n+1} = (2n+1)\pi/(2h), \tag{6.8}$$

where $n = 0, 1, 2, \ldots$. The A_m are constants.

6.1.1 Partial Waves and the Transverse Resonance Principle

The central feature of a waveguide is that the waves phase match in the propagation direction and stand in the transverse direction. Equivalently stated, as the waves reflect back and forth within the guide, they must interfere constructively to reconstruct themselves and form a sustained wavefield. There are two sets of plane waves in a guide, as indicated in Fig. 6.1 – one propagating downward and one upward. The members of each set are referred to as partial waves, just as were the individual plane waves in each cell of the periodic structure examined in Section 1.4. Those that propagate downward are indicated by the solid lines and those that propagate upward by the dashed ones. The downward and

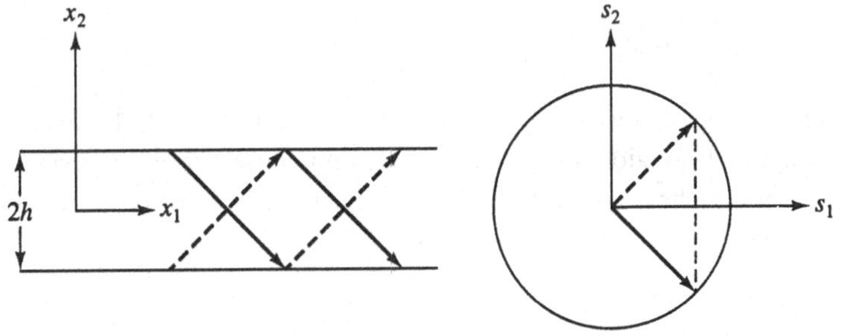

Fig. 6.1. The dashed lines indicate the upward propagating set of plane waves, while the solid lines indicate the downward propagating set. These rays are shown in the slowness diagram to the right. Note that phase matching must occur in the x_1 direction.

upward propagating sets are

$$u_3 = Ae^{i(\beta x_1 - \gamma x_2)}, \tag{6.9}$$

$$u_3 = Be^{i(\beta x_1 + \gamma x_2)}, \tag{6.10}$$

respectively. The term γ is given by (6.6). We have, at this stage, assumed only that $\Re(\gamma) \geq 0$. From our previous work, given in Section 3.2, we can, by a slight modification of that calculation, show that the reflection coefficient at $x_2 = \pm h$ is one. Each upward propagating wave must be reflected into a downward propagating one and vice versa. Therefore, at the upper and lower boundaries,

$$\frac{Ae^{-i\gamma h}}{Be^{i\gamma h}} = 1, \tag{6.11}$$

$$\frac{Be^{-i\gamma h}}{Ae^{i\gamma h}} = 1, \tag{6.12}$$

respectively. The terms A and B are constants. For these two equations to hold simultaneously, the following must be true:

$$\gamma h = m\pi/2, \tag{6.13}$$

and

$$A = B \qquad \text{for } m = 2n, \quad n = 0, 1, 2, \ldots \tag{6.14}$$

or

$$A = -B \qquad \text{for } m = 2n+1, \quad n = 0, 1, 2, \ldots. \tag{6.15}$$

Therefore

$$u_{3m} = A_m \begin{matrix}\cos\\ \sin\end{matrix}(\gamma_m x_2)e^{i\beta_m x_1}, \quad m = 0, 1, 2, \ldots, \tag{6.16}$$

where β_m is given by (6.4), $\gamma_m = m\pi/2h$, and A_m is $2A$ or $2iA$. This method of finding the dispersion relation, (6.4), wherein partial waves are made to stand or resonate in the transverse direction, is referred to as the *transverse resonance principle*.

6.1.2 Dispersion Relation: A Closed Waveguide

The most intriguing feature of guided waves is that the lateral wavenumber β_m is a function of ω, or, alternatively ω is a function of β_m. The dispersion relation

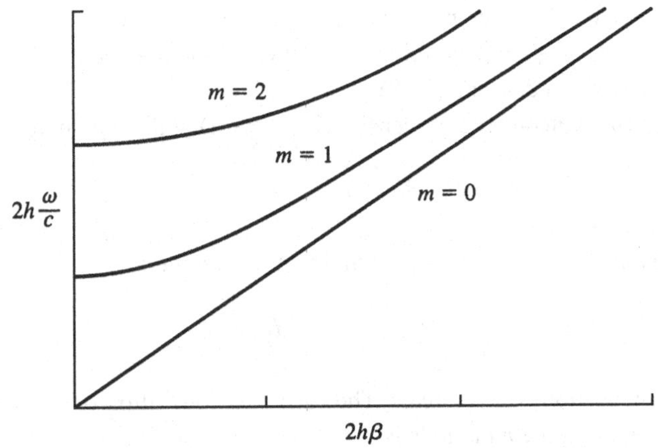

Fig. 6.2. A sketch of the dispersion relation for antiplane shear modes in a closed waveguide. The normalized angular frequency is plotted as a function of the normalized lateral wavenumber.

can be written as

$$2h\beta_m = [(2h\omega/c)^2 - (m\pi)^2]^{1/2}. \tag{6.17}$$

A sketch of this relation is shown in Fig. 6.2. While the β_m may be thought of as a function of ω, when the frequency is a free variable, it is usual to consider ω as a function of β_m. For the mth mode, there is no propagation for $(2h\omega/c) \leq m\pi$ and the frequency $\omega_m = m\pi c/2h$ is called the cut-off frequency. The phase velocity c_m for the mth mode is

$$c_m = \omega/\beta_m. \tag{6.18}$$

The velocity of real interest, however, is the group velocity C_m. Unlike c_m, C_m for the mth mode is given by the slope of the mth branch of the dispersion relation, namely

$$C_m = d\omega/d\beta_m. \tag{6.19}$$

The group velocity is both the velocity of energy propagation and the velocity with which information propagates. Here we demonstrate the first assertion and leave to Section 6.6 the demonstration of the second. The time average of a wavefield quantity $\mathcal{G}_m(\beta_m x_1 - \omega t, x_2)$ for the mth waveguide mode is defined as

$$\langle \mathcal{G}_m \rangle = \frac{1}{T}\int_t^{t+T}\int_{-h}^{h} \mathcal{G}_m(\beta_m x_1 - \omega\tau, x_2)dx_2 d\tau. \tag{6.20}$$

The average $\langle \mathcal{G}_m \rangle$ does not depend upon x_1 or t because at a given x_1, the argument $\beta_m x_1 - \omega\tau$ is simply a translation of the argument $\omega\tau$, a fact noted previously when deriving (2.18).

The instantaneous energy density $\mathcal{E}_m(x_1, x_2, t)$ in the mth mode is

$$\mathcal{E}_m = \tfrac{1}{2}\rho\, \partial_t u_{3m} \partial_t u_{3m} + \tfrac{1}{2}\mu\, \partial_a u_{3m} \partial_a u_{3m}, \tag{6.21}$$

and, using (6.20) with (2.18), we find that its average is

$$\langle \mathcal{E}_m \rangle = \tfrac{1}{2}\rho c_m^2 h \beta_m^2 A_m A_m^*. \tag{6.22}$$

For propagation β_m is real. The instantaneous flux of energy density $\mathcal{F}_m(x_1, x_2, t)$ in the mth mode is

$$\mathcal{F}_m = -\mu \partial_1 u_{3m} \partial_t u_{3m}, \tag{6.23}$$

and the average flux is

$$\langle \mathcal{F}_m \rangle = \tfrac{1}{2}\mu c_m h \beta_m^2 A_m A_m^*. \tag{6.24}$$

The velocity of energy propagation along the axis of the waveguide for the mth mode is, therefore, the ratio $\langle \mathcal{F}_m \rangle / \langle \mathcal{E}_m \rangle$. Direct calculation shows that

$$\langle \mathcal{F}_m \rangle / \langle \mathcal{E}_m \rangle = c^2 / c_m = C_m. \tag{6.25}$$

The average (6.20) can also be used to demonstrate the equipartition of energy. Using this principle when calculating the average energy density, instead of proceeding directly as we just did, usually reduces the amount of calculation needed, though care is required because not all equations describing wavefields exhibit equipartition. The Lagrangian density $\mathcal{L}$ for antiplane shear motion is given by

$$\mathcal{L} = \left(\tfrac{1}{2}\right) \rho (\partial_t u_3)^2 - \left(\tfrac{1}{2}\right) \mu [(\partial_1 u_3)^2 + (\partial_2 u_3)^2]. \tag{6.26}$$

Using (6.20) we readily find that $\langle \mathcal{L} \rangle = 0$, for the mth mode. It follows then that $\langle \mathcal{K} \rangle = \langle \mathcal{U} \rangle$, where the kinetic and internal energy densities, $\mathcal{K}$ and $\mathcal{U}$, were first defined by (1.25). Clearly, the converse is also true.

Dispersion need not arise from a geometrical constraint, as it does with a waveguide, but can arise from the structure of the equation itself, as the following problem demonstrates.

Problem 6.2 Dispersion 1

Problem 1. Consider the following two differential equations, the Klein–Gordon equation,

$$\partial_t^2\varphi - a^2\partial_x^2\varphi + b^2\varphi = 0, \tag{6.27}$$

and the equation for flexural motion in a rod,

$$\partial_t^2\varphi + a^2\partial_x^4\varphi = 0. \tag{6.28}$$

The terms a and b are constants. In both cases find a dispersion relation in the form $\omega = \omega(k)$ by seeking a wave solution of the form

$$\varphi = Ae^{i(kx-\omega t)}, \tag{6.29}$$

where A is a constant. Assume that the wave propagates to the right. Calculate the phase and group velocities c and C, where $c = \omega/k$ and $C = d\omega/dk$.

Demonstrate that the group velocity is the velocity of time-averaged energy transport. To do so you will need to know both the instantaneous energy density and instantaneous flux, so that you can calculate their average values. One way to find these is to construct conservation laws directly from (6.27) and (6.28). Such laws have the form

$$\partial_t\square + \partial_x\square = -\square. \tag{6.30}$$

The first box is the instantaneous energy density, the second the instantaneous flux, and the third, on the right-hand side, a dissipative term. For the equations being studied here, the right-hand side is zero. To find a conservation law, multiply each of (6.27) and (6.28) by $\partial_t\varphi$ and configure the result to coincide with (6.30). Thus, show that the instantaneous energy density for the Klein–Gordon equation is

$$\mathcal{E} = (\partial_t\varphi)^2/2 + a^2(\partial_x\varphi)^2/2 + b^2\varphi^2/2 \tag{6.31}$$

and the instantaneous flux is

$$\mathcal{F} = -a^2(\partial_t\varphi)(\partial_x\varphi). \tag{6.32}$$

Also, show that the instantaneous energy density for the equation for flexural motion is

$$\mathcal{E} = (\partial_t\varphi)^2/2 + a^2\left(\partial_x^2\varphi\right)^2/2 \tag{6.33}$$

and the instantaneous flux is

$$\mathcal{F} = a^2(\partial_t\varphi)(\partial_x^3\varphi) - a^2(\partial_t\partial_x\varphi)(\partial_x^2\varphi). \tag{6.34}$$

Using (2.18), complete the demonstration.

Problem 2. The equation describing acoustic waves in a wind having speed $U\,(<c$, where c is the speed of sound) in the x_1 direction is given by

$$(\partial_t + U\partial_1)^2\varphi = c^2\nabla^2\varphi, \tag{6.35}$$

where $(\boldsymbol{x}, t)$ is a fixed coordinate system. Using a solution of the form

$$\varphi = Ae^{i(\boldsymbol{k}\cdot\boldsymbol{x}-\omega t)}, \tag{6.36}$$

find the dispersion relation. The vector $\boldsymbol{k}$ is needed because the propagation environment is anisotropic.

6.2 Harmonic Waves in an Open Waveguide

We continue to consider an antiplane problem having the governing equation, (6.1). We now consider a layer on a half-space. Figure 6.3 indicates the geometry. Note that the positive x_2 direction points into the interior. The interior of the

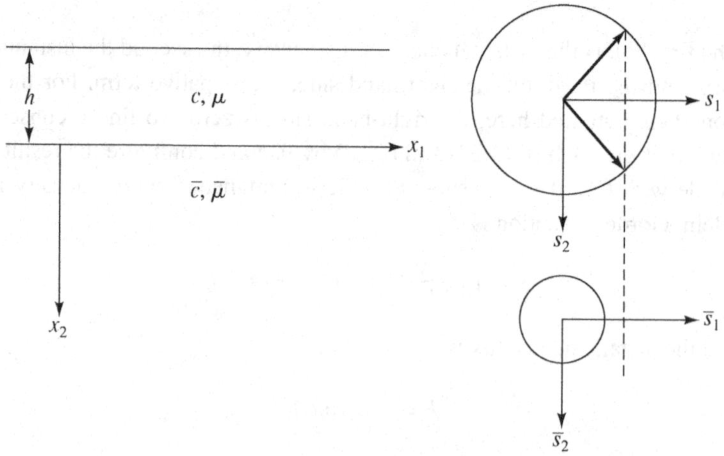

Fig. 6.3. A slow-on-fast structure can support trapped waves in the layer. Drawing slowness diagrams for the layer and half-space indicates the condition for waves to be trapped in the layer, namely that the vertical dashed line in the diagram to the right not intersect the lower slowness circle.

layer occupies $x_2 \in (-h, 0)$ and its wavespeed is c $(k = \omega/c)$. The equation of motion in the layer is (6.1). The interior of the half-space occupies $x_2 \in (0, \infty)$. The equation of motion in the half-space is also given by (6.1) with a different wavenumber. The equation, rewriting it once again, is

$$\partial_\alpha \partial_\alpha u_3 + \bar{k}^2 u_3 = 0, \tag{6.37}$$

where $\bar{k} = \omega/\bar{c}$ and $\bar{c}$ is the wavespeed in the half-space. At $x_2 = -h$,

$$\mu \partial_2 u_3(x_1, -h) = 0, \tag{6.38}$$

and at $x_2 = 0$,

$$u_3(x_1 0^-) = u_3(x_1 0^+), \qquad \mu \partial_2 u_3(x_1, 0^-) = \bar{\mu} \partial_2 u_3(x_1, 0^+), \tag{6.39}$$

where μ and $\bar{\mu}$ are the elastic constants for the layer and half-space, respectively.

The layer is called an open waveguide because there is now the possibility that waves may not remain trapped, but may radiate into the half-space. However, waves will remain trapped provided the wavespeed in the layer is less than that in the half-space. This configuration is sometimes referred to as a slow-on-fast guiding structure. To understand why waves are trapped, consider the slowness diagrams sketched in Fig. 6.3. The wavespeed in the layer is c and that in the underlying half-space $\bar{c}$, so that the corresponding slownesses are s and $\bar{s}$, respectively. A slow-on-fast structure is one for which $s > \bar{s}$. If the waves are to phase match at $x_2 = 0$, s_1 must be the same in both the layer and half-space. If the waves are to remain trapped in the layer, unable to radiate away, $\bar{s}_2$ must be imaginary, with the sign of $\bar{s}_2$ selected so that decay takes place with depth. Therefore, the waves in the layer able to phase match to waves decaying into the half-space are those for which $s_1 = \bar{s}_1 > \bar{s}$. When $s_1 \leq \bar{s} < s$, waves launched in the layer phase match to waves in the half-space for which $\bar{s}_2$ is real and are therefore not trapped. The slow-on-fast structure is analyzed next by using partial waves and the transverse resonance principle. In *Problem 6.3* the problem is reduced to solving an eigenvalue problem in the x_2 direction. A fast-on-slow structure ($s < \bar{s}$) does not guide waves so that, except in *Problem 6.4*, we do not consider it.

6.2.1 Partial Wave Analysis

In the layer, the upward and downward propagating sets of waves are

$$u_3 = B e^{i(\beta x_1 - \gamma x_2)}, \tag{6.40}$$

$$u_3 = A e^{i(\beta x_1 + \gamma x_2)}. \tag{6.41}$$

In the half-space the set of waves is

$$u_3 = \bar{A}e^{i(\beta x_1 + \bar{\gamma} x_2)}. \tag{6.42}$$

The waves in the half-space decay or propagate downward. This statement is consistent with the principle of limiting absorption, given in Section 4.4. To satisfy the equations of motion,

$$\gamma = (k^2 - \beta^2)^{1/2}, \qquad \bar{\gamma} = (\bar{k}^2 - \beta^2)^{1/2}, \tag{6.43}$$

where $\Re(\gamma) \geq 0$ and $\Im(\bar{\gamma}) \geq 0$. We discuss the branches of these radicals more carefully in Section 6.4. The reflection and transmission coefficients at $x_2 = 0$ are, respectively,

$$R(\gamma) = (\mu\gamma - \bar{\mu}\bar{\gamma})/(\mu\gamma + \bar{\mu}\bar{\gamma}), \tag{6.44}$$

$$T(\gamma) = (2\mu\gamma)/(\mu\gamma + \bar{\mu}\bar{\gamma}). \tag{6.45}$$

The $R(\gamma)$ and $T(\gamma)$ are gotten from (3.21) and (3.22) by noting that $\beta = k \sin\theta_0 = \bar{k}\sin\theta_2$, $\gamma = k\cos\theta_0$, and $\bar{\gamma} = \bar{k}\cos\theta_2$. Exactly as in the case of the closed guide, at $x_2 = -h$,

$$Ae^{-i\gamma h}/Be^{i\gamma h} = 1, \tag{6.46}$$

and at $x_2 = 0$,

$$B/A = (\mu\gamma - \bar{\mu}\bar{\gamma})/(\mu\gamma + \bar{\mu}\bar{\gamma}). \tag{6.47}$$

Moreover, in the case of the open guide, we also have

$$\bar{A}/A = (2\mu\gamma)/(\mu\gamma + \bar{\mu}\bar{\gamma}). \tag{6.48}$$

These last two equations indicate that plane waves incident on the lower boundary must, in addition to being reflected, be transmitted into those that decay or propagate away from the boundary. From (6.46) and (6.47) we find that for nontrivial solutions,

$$e^{-2i\gamma h} = (\mu\gamma - \bar{\mu}\bar{\gamma})/(\mu\gamma + \bar{\mu}\bar{\gamma}). \tag{6.49}$$

The case of interest is $\bar{s} < s_1 \leq s$, where $\beta = \omega s_1$, so that $\bar{\gamma} = i\bar{\alpha}$, where $\bar{\alpha} \geq 0$. In this case (6.49) reduces to

$$\tan\gamma h = \bar{\mu}\bar{\alpha}/\mu\gamma. \tag{6.50}$$

Recalling (6.43), we see that (6.49) or (6.50) is the dispersion relation for the structure. It gives $\omega = \omega(\beta)$ or $\beta = \beta(\omega)$, albeit indirectly. The antiplane waves guided by the layer are called *Love waves.* Knowing the dispersion relation, we can find B and $\bar{A}$ in terms of A, and thus construct the wavefields in the layer and half-space. Note that the absence of a plane of reflection symmetry means that the modes do not divide into symmetric and antisymmetric ones.

Problem 6.3 A Second Eigenvalue Problem: An Open Waveguide

Note that (6.29) and (6.30), with boundary conditions (6.31) and (6.32), can be reduced to a singular eigenvalue problem (Friedman, 1956) in x_2 by looking for solutions of the form

$$u_3 = f(x_2)e^{i\beta x_1}, \qquad x_2 \in (-h, 0), \tag{6.51}$$

$$u_3 = g(x_2)e^{i\beta x_1}, \qquad x_2 \in (0, \infty). \tag{6.52}$$

Note that the form of the function in x_1 has been chosen to give a wave propagating in the positive x_1 direction and ensure phase matching at $x_2 = 0$. Use the condition that $g(x_2) \to 0$ as $x_2 \to \infty$. Show that solutions to the differential equation in x_2 and its boundary conditions can be expressed as

$$f(x_2) = C\cos[\gamma(x_2 + h)], \qquad g(x_2) = De^{-\bar{\alpha}x_2}, \tag{6.53}$$

where $\bar{\gamma} = i\bar{\alpha}, \bar{\alpha} \geq 0$. The transverse wavenumbers γ and $\bar{\gamma}$ are given by (6.36). Find the constant D in terms of C. Can you recover the dispersion relationship, (6.50)?

The problem just discussed is called a singular eigenvalue problem because it has a continuous spectrum as well as a discrete one. The preceding analysis gives the discrete eigenvalues and eigenfunctions, but gives one little information about the continuous spectrum, unless the reader is quite perceptive. Friedman (1956), mentioned previously, will help the reader understand the underlying mathematics of singular eigenvalue problems; Tolstoy and Clay (1987) will give the reader a reasoned physical explanation for the continuous spectrum. What condition stated previously in this problem must be modified to find the continuous eigenvalues and eigenfunctions?

6.2.2 Dispersion Relation: An Open Waveguide

The transcendental equation, (6.50), in combination with (6.43), gives either $\omega = \omega(\beta)$ or $\beta = \beta(\omega)$. To examine this dispersion relation, we begin with a diagram such as that shown in Fig. 6.4, where three distinct regions are identified. First we identify the boundaries between which β is real (assuming that $c < \bar{c}$). These boundaries are denoted by c and $\bar{c}$, though the slopes are 1 and $\bar{c}/c$.

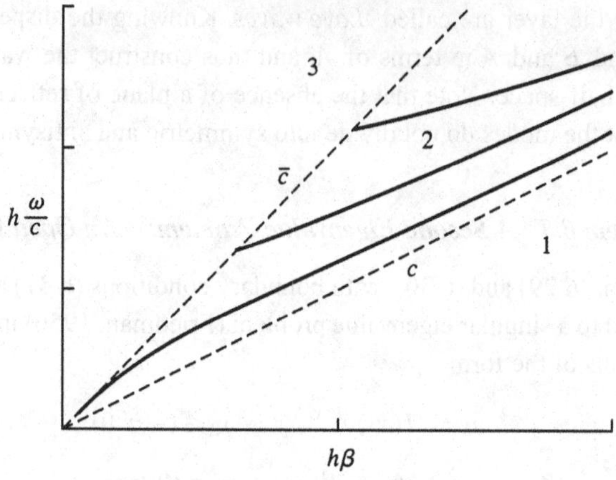

Fig. 6.4. A sketch of the dispersion relation for Love waves. The normalized angular frequency is plotted as a function of the normalized lateral wavenumber. There are three regions whose boundaries are formed by the two wavespeeds $c < \bar{c}$.

In region 1, $\gamma = \pm i\alpha$ and $\bar{\gamma} = i\bar{\alpha}$, where α and $\bar{\alpha}$ are real and positive. It is readily seen that no solution for real β is possible and therefore no trapped wave propagates in the x_1 direction. In region 3 both γ and $\bar{\gamma}$ are real. In this case (6.43) becomes

$$\tan(\gamma h) = -i\bar{\mu}\bar{\gamma}/\mu\gamma. \tag{6.54}$$

Again it is clear that there is no solution for β that is real, though there are solutions for complex β. We shall briefly discuss this possibility in Section 6.4.3. The absence of roots for β real tells us that there are no trapped waves, exactly what we would expect for γ and $\bar{\gamma}$ both real.

Solutions for real β are possible in region 2, as can be seen from examining (6.50). Consider the following limits. First, let $\beta h \to \infty$ within region 2. Write (6.50) as

$$\tan\left\{\beta h[(k^2/\beta^2) - 1]^{1/2}\right\} = \frac{\bar{\mu}[1 - (\bar{k}^2/\beta^2)]^{1/2}}{\mu[(k^2/\beta^2) - 1]^{1/2}}. \tag{6.55}$$

The argument of the tangent in (6.55) approaches a finite limit or zero because the right side remains positive, and approaches zero or $+\infty$. There is only one possibility, namely $k/\beta \to 1$. In other words, by allowing the layer to become an infinite number of wavelengths thick, its response is no longer influenced by the presence of the half-space. Second, let βh move toward zero through region 2 and reason as before. For the lowest mode $\bar{k}/\beta \to 1$ as $\beta h \to 0$ and

$\omega h \to 0$. However, for the higher modes $\bar{k}/\beta \to 1$, while βh and ωh remain finite. In this case, for some $\omega = \omega_n$, $\bar{\alpha} = 0$ and

$$[(\omega_n h/\beta hc)^2 - 1]^{1/2} \beta h = n\pi, \tag{6.56}$$

where n is a positive integer. The frequency ω_n is that at which the waves move from being trapped to radiating into the half-space. This frequency is also called a cut-off frequency, though the term *transition frequency* might be more appropriate. Unlike the case of a closed waveguide, the group velocity does not vanish at this point. Keep in mind that we have, so far, restricted β to positive real values and have not defined the branches of γ and $\bar{\gamma}$ carefully.

The outcomes of the problem just described should be compared with those of the reflection problem that follows. The viewpoint there is rather different. In the reflection problem the layer is viewed from the half-space, and we are no longer as concerned as we were here with whether the structure is a slow-on-fast or fast-on-slow one.

Problem 6.4 Reflection From a Layer

Referring to the geometry of Fig. 6.3, let the plane wave

$$\bar{u}_{30} = A_0 e^{i(\beta x_1 - \bar{\gamma} x_2)} \tag{6.57}$$

be incident to the layer from the half-space. Make no assumption at present as to the relative magnitudes of c and $\bar{c}$. If A_1 is the unknown amplitude of the reflected wave, calculate the reflection coefficient of the layer, namely $R(\theta_0) = A_1/A_0$. Show that $R(\theta_0) = A_-(\theta_0)/A_+(\theta_0)$, where

$$A_\mp = \cos\theta_0 \cos(kh\cos\theta) \mp i\frac{\mu\bar{c}}{\bar{\mu}c}\cos\theta \sin(kh\cos\theta), \tag{6.58}$$

and

$$\sin\theta/c = \sin\theta_0/\bar{c}. \tag{6.59}$$

The wavenumbers $\beta = \bar{k}\sin\theta_0$ and $\bar{\gamma} = \bar{k}\cos\theta_0$.

The reader should contrast the two possible cases, fast-on-slow and slow-on-fast. In particular, how might a trapped wave be excited by using a disturbance incident from the half-space when the structure is slow on fast? Note that this reflection coefficient is frequency dependent.

6.3 Excitation of a Closed Waveguide

6.3.1 Harmonic Excitation

We start by considering a closed waveguide that is harmonically excited by applying an antiplane traction at one end, while the other extends to infinity. One common way to solve such problems is to expand the wavefield in the guide in terms of its modes. To do so requires that a mathematical framework be in place. As a minimum we need a means of calculating the coefficients of the expansion – an orthogonality relation – and some assurance that the expansion is complete. While these questions have still not been satisfactorily answered for inplane elastic waves in a guide with an end (Miklowitz, 1978; Folguera and Harris, 1999), the antiplane modes are simply the terms of a Fourier series, so that in this case these questions are settled.

Problem 6.5 A Waveguide Mode Expansion

In part, (6.1) and (6.2) specify the problem. The geometry of the problem is now described by a layer similar to that shown in Fig. 6.1, but starting at $x_1 = 0$ and extending through positive values of x_1 to infinity. At $x_1 = 0$ the waveguide is excited with a traction $T(x_2)$ symmetric in x_2 so that

$$\mu \partial_1 u_3 = -T(x_2), \qquad T(x_2) = T(-x_2). \tag{6.60}$$

What condition should be imposed upon the forced disturbance as $x_1 \to \infty$? Noting the symmetry of the excitation, how might the forced wavefield be expanded? Show that a representation of the solution is

$$u_3(x_1, x_2) = \sum_{n=0}^{N-1} \frac{(-A_n)}{i\beta_n} \cos\left(\frac{n\pi}{h} x_2\right) e^{i\beta_n x_1} + \sum_{n=N}^{\infty} \frac{A_n}{\alpha_n} \cos\left(\frac{n\pi}{n} x_2\right) e^{-\alpha_n x_1}, \tag{6.61}$$

where N is the smallest n for which the inequality

$$n\pi/h < \omega/c \tag{6.62}$$

is reversed. The frequency of excitation is ω, β_n is given by (6.4), and, for $n \geq N$, $\beta_n = i\alpha_n$. Using the orthogonality of the modes (eigenfunctions), write down explicit expressions for the coefficients A_n.

Note that the modes in the first sum of (6.61) propagate, while those in the second are evanescent. Far from the source, only those that propagate make a significant contribution.

6.3.2 Transient Excitation

We next consider a transient excitation. In *Problem 6.5*, we replace (6.60) with

$$\mu\partial_1 u_3 = -T(x_2)\delta(t). \tag{6.63}$$

Moreover, we assume that the waveguide is quiescent until the excitation is applied. That is,

$$u_3 = \partial_t u_3 = 0, \qquad t < 0. \tag{6.64}$$

Using (1.36), we synthesize the transient response as

$$u_3(x_1, x_2, t) = \sum_{n=0}^{\infty} \frac{(-A_n)}{\pi} \cos\left(\frac{n\pi}{h} x_2\right) \Re \int_0^{\infty} \frac{e^{it(\beta_n x_1/t - \omega)}}{i\beta_n} d\omega. \tag{6.65}$$

Using the principle of limiting absorption, replace ω with $\omega + i\epsilon$, where $\epsilon > 0$. The $n = 0$ term is easily inverted, giving

$$\Re \int_0^{\infty} \frac{e^{i(\omega+i\epsilon)(x_1/c-t)}}{i(\omega+i\epsilon)/c} d\omega = -\pi c\, H(t - x_1/c). \tag{6.66}$$

The $n > 0$ terms are less straightforward. The integral to be evaluated is

$$\Re \int_0^{\infty} \frac{e^{ix_1(\beta_n - \omega t/x_1)}}{i\beta_n} d\omega. \tag{6.67}$$

The left half of Fig. 6.5 shows the complex ω plane and the branch cuts for $n > 0$. Following the arguments in Section 3.4.4, we have cut the plane so that $\Re(\beta_n) \geq 0$ for all ω. Using the principle of limiting absorption for $\omega > 0$, we find the branch point is at $cn\pi/h - i\epsilon$. By symmetry, for $\omega < 0$, the branch point is at $-cn\pi/h + i\epsilon$. The integration contour in the ω plane must pass above the branch point $cn\pi/h - i\epsilon$. Note that $cn\pi/h$ is a cut-off frequency so that propagation has ceased when $\omega < cn\pi/h$. Note, as well, that $\Im(\beta_n) > 0$ in the first and third quadrants. Once the contour has been determined, the ϵ in (6.67) can be set to zero.

Integrals of the form of (6.67), even when they can be evaluated in closed form, give the clearest description of the wavefield at points far from the source. Accordingly, we approximate (6.67) by using the method of stationary phase described in Section 5.3.3. We fix t/x_1 and let x_1 be the large parameter. For a

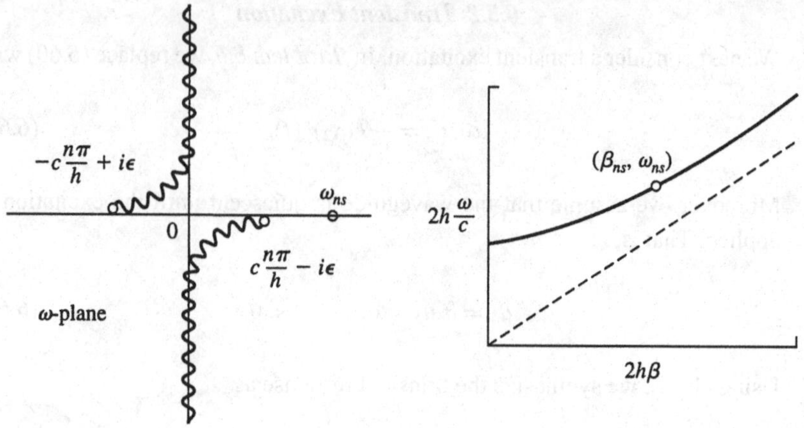

Fig. 6.5. The left-hand figure shows the complex ω plane. The contour extends from $\omega = 0$ along the real axis. $\Re(\beta_n) \geq 0 \,\forall\, \omega$, $\Im(\beta_n) > 0$ in the upper right quadrant of the ω plane. The right-hand figure shows the dispersion curve for the nth mode with the stationary point indicated by $(\beta_{ns}, \omega_{ns})$.

given point in (x_1, t), the stationary point is given by

$$d\beta_n/d\omega = t/x_1. \tag{6.68}$$

Solving this equation gives the stationary point $(\beta_{ns}, \omega_{ns})$ as a function of (x_1, t) (more precisely x_1/t), namely

$$\frac{\omega_{ns}}{c} = \frac{n\pi/h}{[1-(x_1/ct)^2]^{1/2}}, \quad x_1/ct < 1, \tag{6.69}$$

and

$$\beta_{ns} = \left(\frac{n\pi}{h}\right)\frac{x_1/ct}{[1-(x_1/ct)^2]^{1/2}}, \quad x_1/ct < 1. \tag{6.70}$$

No signal can travel faster than c (*Problem 6.6*). Therefore, $x_1/ct < 1$ for propagating waves and $(\beta_{ns}, \omega_{ns})$ are real.

The integral, (6.67), is then approximated as

$$\int_0^\infty \frac{e^{ix_1(\beta_n - \omega t/x_1)}}{i\beta_n}\,d\omega \sim \frac{c(2\pi)^{1/2}}{(n\pi/h)(\beta_{ns}x_1)^{1/2}}\,e^{ix_1(\beta_{ns}-\omega_{ns}t/x_1)}e^{-i\pi/4}, \tag{6.71}$$

as $x_1 \to \infty$, while t/x_1 is held fixed. After some simplification, the particle displacement of the nth mode is approximated as

$$u_{3n}(x_1, x_2, t) \sim \frac{(-A_n)}{\pi} c \cos\left(\frac{n\pi}{h} x_2\right) \left\{ \frac{(2h/n)}{x_1[(ct/x_1)^2 - 1]^{1/2}} \right\}^{1/2}$$
$$\times \cos\left\{(n\pi/h) x_1 [(ct/x_1)^2 - 1)]^{1/2} + \pi/4\right\}, \qquad (6.72)$$

and the total particle displacement can be written as

$$u_3(x_1, x_2, t) \sim A_0 c\, H(t - x_1/c) + \sum_{n=1}^{\infty} u_{3n}(x_1, x_2, t). \qquad (6.73)$$

Examining (6.72) shows that the nth mode is $O(n^{-1/2})$. The higher-order modes therefore make a weaker contribution than do the lower-order ones.

Note that the stationary phase approximation breaks down for x_1/t near c. Dai and Wong (1994) give a correction for this case. Further, note that as x_1/t sweeps through its values from zero to c, ω_s takes on all its values from $cn\pi/h$ to infinity. In fact, for fixed x_1, (6.68) provides a one-to-one mapping from the t plane to the ω plane.

At (x_1, t) the nth mode is approximated by an expression of the form

$$\Re\left[A_n(x_1, x_2, t) e^{i(\beta_{ns} x_1 - \omega_{ns} t)}\right]. \qquad (6.74)$$

This is a group in the sense that it arises from a cluster or group of wavenumbers and frequencies in the neighborhood of $(\beta_{ns}, \omega_{ns})$. Note that (6.68) can be viewed as $x_1/t = C_n(\omega_{ns})$, where C_n is the group velocity $d\omega/d\beta_n$. We have already noted in (6.25) that C_n is the velocity of time-averaged energy transmission. Now we also note that it is the velocity with which the frequency ω_{ns} propagates to the point (x_1, t). This is not the velocity with which the phase $\theta(x_1, t) = (\beta_{ns} x_1 - \omega_{ns} t)$ propagates outward. Rather, a constant phase θ propagates at a rate $c_{ns} = \omega_{ns}/\beta_{ns}$.

Problem 6.6 No Signal Travels Faster than the Velocity c

Using the problem just discussed, show that no signal travels faster than c. Write each integral over ω as

$$\frac{\Re}{\pi} \int_0^{\infty} e^{-i\omega(t - x_1/c)} \frac{e^{i(\beta_n x_1 - \omega x_1/c)}}{i\beta_n} d\omega \qquad (6.75)$$

and note that, in the upper right quadrant of Fig. 6.5, with the ω plane cut as indicated,

$$\lim_{|\omega|\to\infty} \frac{e^{i(\beta_n x_1 - \omega x_1/c)}}{i\beta_n} = 0. \tag{6.76}$$

Therefore, for $t < x_1/c$, show that the integration contour may be distorted from one along the real ω axis to the one along the imaginary axis. Lastly, show that the integral can be written as

$$\frac{\Re}{\pi} \int_0^{\infty} i e^{p(t - x_1/c)} \frac{e^{-(\alpha_n x_1 - p x_1/c)}}{(-\alpha_n)} dp. \tag{6.77}$$

This integral must be zero (why?), indicating that no signal is present for $t < x_1/c$.

While this result is the outcome of the analysis of a specific problem, it is true generally. Moreover, it indicates that the group velocity does not exceed the wavespeed of the medium,[2] a result that is also generally true.

For separable problems, such as the one discussed in *Problem 6.5*, an expansion of the solution in the eigenfunctions of the transverse eigenvalue problem results, after invoking the orthogonality of the eigenfunctions, in reducing the partial differential equation to an ordinary one in x_1, involving each mode separately. The domain of the transverse eigenvalue problem is finite and therefore the eigenvalues are discrete. Expanding the source in these eigenfunctions causes the equation in x_1 to be forced by a coefficient of this expansion. This is more or less what we did in solving *Problem 6.5*, though the modal expansion already contained the solution to the differential equation in x_1, so that that step was leaped over. In *Problem 6.7* the reader is asked to solve much the same problem by using a continuous eigenfunction expansion in $x_1 \in (0, \infty)$, thus reducing the partial differential equation to an ordinary one in x_2. Moreover, rather than use a Fourier synthesis to construct the transient solution, the reader is asked to use a Laplace transform.

[2] The dispersion is said to be anomalous when the group velocity exceeds the velocity of propagation of a harmonic plane wave in the medium (Sommerfeld, 1964a; Brillouin, 1960). In this case the group velocity is no longer a measure of the speed at which information or energy propagates. In such cases a specific signaling problem must be worked through, and velocities that characterize the speed at which information is propagated and at which energy is propagated must be defined as best one can. Moreover, the anomalous dispersion may only be apparent and not real. One must take care to distinguish between anomalous dispersion caused by some approximate derivation of the equation being studied and that caused by the underlying physical situation. The presence of frequency-dependent attenuation confuses the issue even further.

Problem 6.7 A Continuous Eigenfunction Expansion

Continue to consider the problem just solved, with the excitation (6.63). Use the cosine transform

$$^{*}u_3(\xi, x_2, t) = \int_0^\infty u_3(x_1, x_2, t)\cos(\xi x_1)dx_1, \tag{6.78}$$

followed by the Laplace transform

$$^{*}\bar{u}_3(\xi, x_2, p) = \int_0^\infty {}^{*}u_3(\xi, x_2, t)e^{pt}dt, \tag{6.79}$$

to find an ordinary differential equation in x_2. Find $^{*}\bar{u}_3$ and begin to invert the transforms. It is perhaps easier to set $^{*}\bar{v}_3(\xi, x_2, p) = p^{*}\bar{u}_3(\xi, x_2, p)$ where v_3 is the particle velocity. Find

$$^{*}v_3(\xi, x_2, t) = \frac{1}{2\pi i}\int_{\epsilon - i\infty}^{\epsilon + i\infty} {}^{*}\bar{v}_3(\xi, x_2, p)e^{pt}dp \tag{6.80}$$

and then approximate

$$v_3(x_1, x_2, t) = \frac{2}{\pi}\int_0^\infty {}^{*}v_3(\xi, x_2, t)\cos(\xi, x_1)d\xi \tag{6.81}$$

for large x_1, with x_1/t held fixed, using the method of stationary phase. This integral can also be inverted exactly.

The reader may be surprised to find that his or her expression for v_3 contains both forward and backward propagating waves. This is because we have chosen to work with a cosine transform. Why would one *not* work with a sine transform? At what point has the reader imposed the condition that waves be outgoing from the source? Can the reader express (6.81) as a sum of waves propagating solely in the positive x_1 direction?

6.4 Harmonically Excited Waves in an Open Waveguide

We next return to the layered antiplane structure, sketched on the left in Fig. 6.3, to calculate the wavefield radiated by a harmonic line source placed at $(0, x_{20})$, where $x_{20} < 0$. Except for the absence of a source, the equations of motion and the boundary conditions are given by (6.38) and (6.39). This problem may be solved by constructing the solution by using eigenfunction expansions. However, as we have done previously, we shall construct the radiated wavefield by superposing collections of partial waves. This method of construction will show how ray representations and modal expansions are related to one another. Our discussion follows a very similar one in Brekhovskikh (1980).

6.4.1 The Wavefield in the Layer

We begin by representing the wavefield radiated by a line source as a spectrum of plane waves, using the result of *Problem 4.1*. Equation (4.19), with α replacing ξ, is here rewritten by expanding $\cos(\theta - \alpha)$ to give

$$u_3^i = -\frac{iF_0}{4\pi} \int_C e^{ik(x_1 \cos\alpha + |x_2 - x_{20}| \sin\alpha)}\, d\alpha. \tag{6.82}$$

$F_0 = A/k$, where A is a dimensionless constant. The expression (6.82) gives the free space, particle displacement u_3^i excited by a line source at $(0, x_{20})$. The contour C begins near $\pi - i\infty$ and ends near $i\infty$. How then does each plane wave in the integral, (6.82), behave in the layered environment?

Consider an arbitrary observation point (x_1, x_2) in the layer, where $x_1 > 0$ and $x_2 > x_{20}$. Given that the x_1 dependence must be of the form $e^{ik\cos\alpha\, x_1}$, in both the layer and half-space, to ensure phase matching, we ask what partial waves can reach the point (x_1, x_2) and, when added together, satisfy the boundary conditions. By sketching rays and wavefronts, as suggested in Fig. 6.6, we find that there are four distinct partial waves that reach (x_1, x_2) and that all the subsequent partial waves can be constructed from these four by successively reflecting them in the planes $x_2 = 0$ and $x_2 = -h$.

$$e^{ik[x_1 \cos\alpha + \sin\alpha(x_2 - x_{20})]}, \quad R(\alpha)e^{ik[x_1 \cos\alpha - \sin\alpha(x_2 + x_{20})]},$$
$$e^{ik[x_1 \cos\alpha + \sin\alpha(x_2 + 2h + x_{20})]}, \quad R(\alpha)e^{ik[x_1 \cos\alpha - \sin\alpha(x_2 - 2h - x_{20})]}. \tag{6.83}$$

The first partial wave reaches (x_1, x_2) without reflection, the second after being

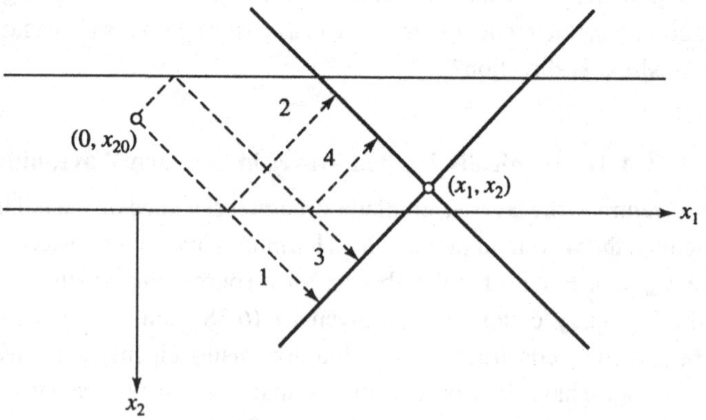

Fig. 6.6. A sketch showing the first four partial waves reaching the observation point (x_1, x_2). The dashed lines indicate the propagation paths and the solid ones indicate the wavefronts passing through (x_1, x_2).

reflected once at the $x_2 = 0$ boundary, the third after being reflected from the $x_2 = -h$ boundary, and the last after being reflected first from the $x_2 = -h$ boundary and second from the $x_2 = 0$ boundary. All subsequent reflected partial waves can be constructed from these four. The reflection coefficient at $x_2 = -h$ is 1 and at $x_2 = 0$ is $R(\alpha)$, where $R(\alpha)$ is given by (6.44), with $\gamma = k \sin\alpha$ and $\bar{\gamma} = \bar{k} \sin\bar{\alpha}$. Or it can be gotten from (3.21) and (3.23) by noting that $\alpha = \pi/2 - \theta_0$. The angles α and $\bar{\alpha}$ are related by the phase-matching condition, namely $k\cos\alpha = \bar{k}\cos\bar{\alpha}$. Note that the argument of the function R is now given as α (rather than as γ or θ_0) in keeping with the structure of the present calculation.

By adding these four partial waves and all the subsequent reflections and invoking superposition, we construct the response of the layer to the excitation. It is given by

$$
\begin{aligned}
u_3 = -\frac{iF_0}{4\pi}\int_C e^{ik\cos\alpha x_1}\Big[e^{ik\sin\alpha(x_2-x_{20})} + R(\alpha)e^{-ik\sin\alpha(x_2+x_{20})} \\
+ e^{ik\sin\alpha(x_2+2h+x_{20})} + R(\alpha)e^{-ik\sin\alpha(x_2-2h-x_{20})}\Big] \\
\times \sum_{m=0}^{\infty}[R(\alpha)]^m e^{ik\sin\alpha(2mh)}\,d\alpha
\end{aligned}
\qquad (6.84)
$$

To understand this representation further, we approximate one of the terms by the method of steepest descents. We consider $m = 0$, the fourth term, namely

$$
u_3^{04} = -\frac{iF_0}{4\pi}\int_C R(\alpha)e^{ik[\cos\alpha\, x_1 - \sin\alpha\,(x_2-2h-x_{20})]}\,d\alpha. \qquad (6.85)
$$

The geometry is shown in Fig. 6.7. By setting $x_1 = r_{04}\cos\theta_{04}$ and $x_2 - 2h - x_{20} = -r_{04}\sin\theta_{04}$, we put (6.85) in a form previously studied in Section 5.4. Its asymptotic approximation is therefore

$$
u_3^{04} = -\frac{iF_0}{4\pi}\int_C R(\alpha)e^{ikr_{04}\cos(\alpha-\theta_{04})}\,d\alpha \sim \left(\frac{2\pi}{kr_{04}}\right)^{1/2}\frac{F_0}{4\pi}R(\theta_{04})e^{ikr_{04}}e^{i\pi/4}. \qquad (6.86)
$$

This term then appears as a cylindrical wave radiated by an image source at $(0, 2h + x_{20})$. The actual path of the ray is indicate by the dashed line within the layer in Fig. 6.7. Approximating each term in (6.84) by the method of steepest descents gives an infinite sum of terms such as (6.86). This is a *ray representation* of the wavefield in the layer.

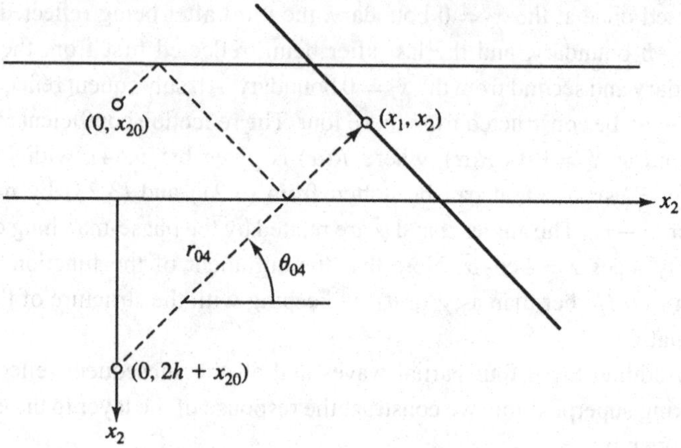

Fig. 6.7. A sketch of the geometry for the steepest descents approximation of the fourth partial wave.

Problem 6.8 Ray Representation

Verify the statements just made by asymptotically approximating each term in (6.84) to give

$$u_3 \sim \sum_{l=0}^{\infty} \sum_{j=1}^{4} u_3^{lj}, \tag{6.87}$$

where each term is similar to that given by (6.86). Is this a useful representation, and, if so, in what circumstances?

By summing the series in (6.84), the wavefield can be recast as

$$u_3 = -\frac{iA}{4k\pi} \int_C e^{ik\cos\alpha\, x_1} \left[e^{ik\sin\alpha\, x_2} + R(\alpha) e^{-ik\sin\alpha\, x_2}\right]$$

$$\times \frac{\left[e^{-ik\sin\alpha\, x_{20}} + e^{ik\sin\alpha\,(2h+x_{20})}\right]}{\left[1 - R(\alpha) e^{ik\sin\alpha(2h)}\right]} d\alpha. \tag{6.88}$$

$F_0 = A/k$ has been used to indicate the dimensions of u_3. Recall that we have assumed $x_2 > x_{20}$. When $x_2 < x_{20}$, the two coordinates in the above expression are interchanged.

To understand this representation we must examine the complex α plane. However, the discussion will be clearer if we first examine the β plane, where $\beta = k\cos\alpha$. Recall that this is the Sommerfeld transformation and that it was this transformation that lead us to (6.82). Figure 6.8 shows the β plane with its branch cuts and several poles. The transverse wavenumbers γ and $\overline{\gamma}$ were given in (6.43). These radicals are defined so that $\Im(\gamma) \geq 0$ and $\Im(\overline{\gamma}) \geq 0 \,\forall\, \beta$.

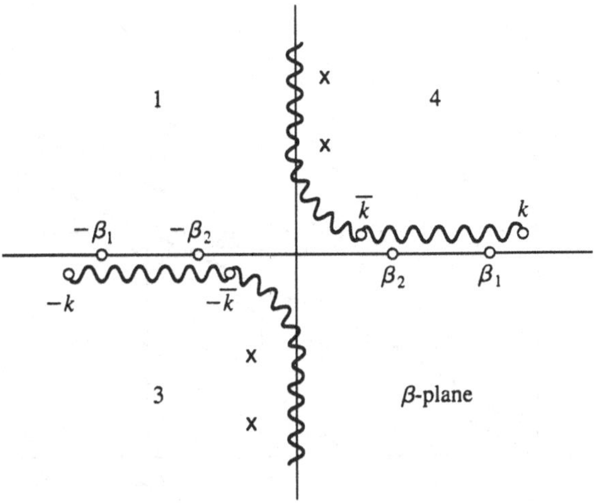

Fig. 6.8. A sketch of the complex β plane, showing the branch cuts and poles that have emerged onto the physical sheet. The physical sheet is the one for which $\Im(\gamma) \geq 0$ and $\Im(\overline{\gamma}) \geq 0 \; \forall \; \beta$. In the various quadrants the real parts take values as follows: quadrants 1 and 2, $\Re(\gamma) > 0$ and $\Re(\overline{\gamma}) > 0$; quadrants 3 and 4, $\Re(\gamma) < 0$ and $\Re(\overline{\gamma}) < 0$. These quadrants correspond to their primed counterparts in the α plane in Fig. 6.11.

This is the Riemann sheet, as we have seen in Sections 3.4.4 and 5.4, that ensures that the waves radiate or decay away from their source (satisfy the principle of limiting absorption). We call this the physical sheet and the others the unphysical ones. Figure 6.9 shows the α plane and the contour C. The branch points are now given by $\overline{\alpha}$ and $\pi - \overline{\alpha}$, where $\overline{\alpha}$ is defined by $\overline{k} = k \cos \overline{\alpha}$.

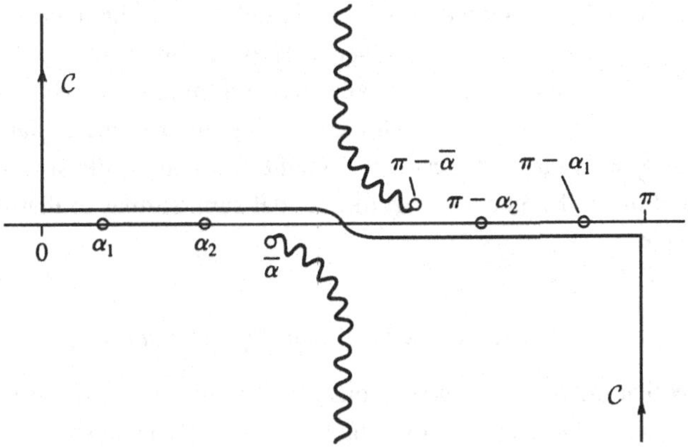

Fig. 6.9. The α plane. The poles to the left are α_1 and α_2; those to the right are $\pi - \alpha_1$ and $\pi - \alpha_2$. The two branch points are $\overline{\alpha}$ and $\pi - \overline{\alpha}$. Poles on the lower sheet progressively move to the physical sheet, popping through the branch points as $\overline{\omega}$ is increased.

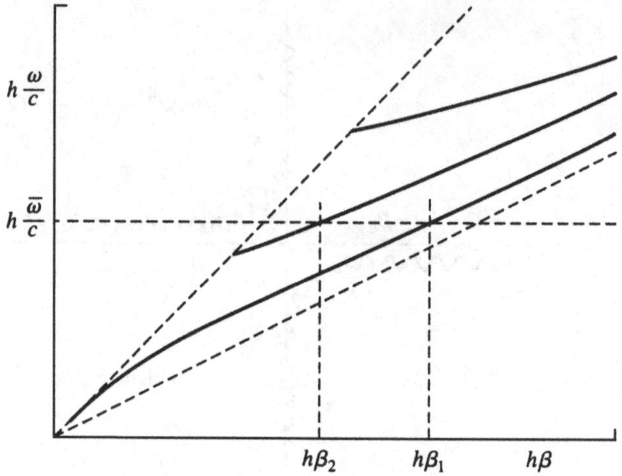

Fig. 6.10. The dispersion relation and the points on it that correspond to the poles shown in Figs. 6.8 and 6.9.

Noting that $\gamma = k \sin\alpha$, we see that *the poles of the integrand in* (6.88) *are given by the dispersion relation*

$$e^{-2i\gamma h} = R(\alpha), \tag{6.89}$$

which is a restatement of (6.49). The poles of the integrand therefore give us the trapped modes of the waveguide. For a given $\omega = \bar{\omega}$, there is a pair of poles on the physical sheet of the β or α plane corresponding to each guided mode that is trapped in the layer. In Figs. 6.8 and 6.9, there are two pairs, $\pm\beta_1$ and $\pm\beta_2$ or $(\alpha_1, \pi - \alpha_1)$ and $(\alpha_2, \pi - \alpha_2)$. These points are shown on the dispersion curves in Fig. 6.10. The modes that could radiate into the interior correspond to poles that lie on an unphysical sheet. These are indicated by the crosses in Fig. 6.8. As $\bar{\omega}$ is increased, these poles move from the unphysical sheet to the physical one, popping up through the branch point $\bar{\alpha}$ in the α plane. The next problem indicates how the residues from the poles on the physical sheet, in the integrand of (6.88), become part of a modal sum, similar to that for a closed waveguide.

Problem 6.9 Modal Representation

Show that the residue contributions of (6.88) give the Love waves of (6.51) through (6.53). Do this by deforming the contour C to that shown in Fig. 6.11. Note that there will be a branch cut integral as indicated in the figure. The sum of residues gives the waves trapped in the layer or the sum over the discrete

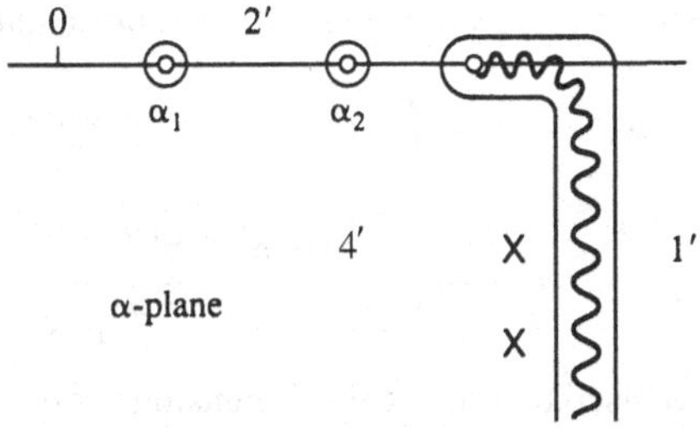

Fig. 6.11. The deformed contour for *Problem 6.9.* Propagation to the right is being considered. The quadrants 1′, 2′, and 4′ correspond to quadrants 1, 2, and 4 in the β plane in Fig. 6.8.

eigenfunctions, whereas the branch cut integral gives the wavefield that can radiate into the lower half-space or the sum over the continuous eigenfunctions. The combination is the modal expansion of the wavefield in the layer.

Ray representations are useful for observation points near the source, whereas modal representations are most descriptive a considerable distance from it, at points where the wavefield has lost some sense of how it was excited and has adapted to its propagation environment. The modal representation can be derived directly from (6.84) by using the Poisson sum formula, discussed in Section 1.3. This is true even when we cannot sum the series (6.84).

6.4.2 The Wavefield in the Half-Space

The wavefield in the half-space is constructed in much the same way as was that in the layer. The two fundamental partial waves reaching (x_1, x_2) are

$$T(\alpha)e^{ik\cos\alpha\, x_1}\, e^{-ik\sin\alpha\, x_{20}}\, e^{i\bar{k}\sin\bar{\alpha}\, x_2}, \qquad T(\alpha)e^{ik\cos\alpha\, x_1}\, e^{ik\sin\alpha\,(x_{20}+2h)}\, e^{i\bar{k}\sin\bar{\alpha}\, x_2}. \tag{6.90}$$

The remainder can be constructed from these two. Integrating over all the partial waves gives

$$u_3 = -\frac{iF_0}{4\pi}\int_C T(\alpha)e^{ik\cos\alpha\, x_1}\, e^{i\bar{k}\sin\bar{\alpha}\, x_2}\left[e^{-ik\sin\alpha\, x_{20}} + e^{ik\sin\alpha\,(2h+x_{20})}\right]$$
$$\times \sum_{m=0}^{\infty}[R(\alpha)]^m e^{ik\sin\alpha\,(2mh)}\, d\alpha. \tag{6.91}$$

If each term were approximated by the method of steepest descents, then this representation would give a ray series for the wavefield in the half-space. By carrying out the summation, we can recast the representation as

$$u_3 = -\frac{iA}{4k\pi} \int_C T(\alpha) e^{ik\cos\alpha\, x_1} e^{i\bar{k}\sin\bar{\alpha}\, x_2}$$

$$\times \frac{\left[e^{-ik\sin\alpha\, x_{20}} + e^{ik\sin\alpha\,(2h+x_{20})}\right]}{\left[1 - R(\alpha) e^{ik\sin\alpha\,(2h)}\right]} d\alpha. \tag{6.92}$$

$F_0 = A/k$ has been used to indicate the dimensions of u_3. The poles and branch cuts of the integrand also give rise to a modal representation.

6.4.3 Leaky Waves

We continue to consider propagation solely to the right. Figures 6.8, 6.9, and 6.11 indicate how the β plane is structured and how it relates to the α plane. Moreover, we continue to assume that there are two real roots, $\beta_1 = k\cos\alpha_1$ and $\beta_2 = k\cos\alpha_2$, present on the physical sheet. To the left of the $\omega/\beta = \bar{c}$ line, in Figs. 6.4 and 6.10, there are also roots to the dispersion relation if β is allowed to be complex. They lie on the $\Im(\gamma) \geq 0$, $\Im(\bar{\gamma}) \leq 0$ sheet, an unphysical sheet of the β plane, and therefore on the unphysical sheet in the α plane. Their positions are approximately indicated by the crosses in both Figs. 6.8 and 6.11. To ensure that waves radiate away from their source, we take $\Im(\beta) > 0$ so that the waves in the layer radiate or leak into the interior as they propagate to the right. We call them leaky waves or leaky modes. Note that $\Re(\bar{\gamma}) > 0$ for propagation to the right, but that $\Im(\bar{\gamma}) < 0$ so that these waves increase exponentially with depth. That is why they do not lie on the physical sheet of the Riemann surface. They can, however, influence the solution if they come close to the branch cut or if the branch cut is moved. These waves can be quite important for some observation points. This issue is discussed at greater length in DeSanto (1992).

6.5 A Laterally Inhomogeneous, Closed Waveguide

We have considered both open and closed waveguides whose geometry has conformed to a rectangular coordinate system. Moreover we have only considered a very simple variation in the propagation environment with depth, namely the layer on a half-space. This last example of waveguiding considers a propagation environment that changes slowly in the propagation or lateral direction. We consider a slow variation, with respect to wavelength, in the thickness of a closed guide whose unperturbed geometry is that given in Fig. 6.1. We construct an asymptotic approximation that makes use of rays to describe the propagation in the lateral direction and modes to take account of the variation in the transverse one.

The equation of motion remains (6.1). To clarify the scales we introduce scaled coordinates $\bar{x}_1 = kx_1$ and $\bar{x}_2 = kx_2$. The slow variable $y_1 = \delta k x_1$ is introduced to describe the slow lateral variation. Also we set $\bar{u}_3 = ku_3$. Having scaled the problem, we *omit the overbar* and reintroduce the variables x_2 and u_3 with the understanding that these are *now scaled variables.* The equation of motion becomes

$$\delta^2(\partial^2 u_3/\partial y_1^2) + (\partial^2 u_3/\partial x_2^2) + u_3 = 0. \tag{6.93}$$

The boundary conditions require the vanishing of the normal traction. This is the second way in which the small parameter δ enters the problem. The top and bottom surfaces are given by

$$x_2 = \pm H^{\pm}(x_1) = \pm h_0 + h^{\pm}(y_1). \tag{6.94}$$

$[H^{\pm}(x_1 + 2\pi) - H^{\pm}(x_1)]/2\pi \approx \delta[dh^{\pm}/dy_1]$, where we assume that $dh^{\pm}/dy_1 = \mathrm{O}(1)$; δ then measures the change in the thickness of the guide over a wavelength. The outward unit normal vectors are

$$\hat{\mathbf{n}}^{\pm} = \left[-\delta \frac{dh^{\pm}}{dy_1}\hat{\mathbf{e}}_1 \pm \hat{\mathbf{e}}_2 + \mathrm{O}(\delta^2) \right]. \tag{6.95}$$

We now seek an asymptotic solution in the form

$$u_3 \sim e^{i\theta(y_1)/\delta} \sum_{\nu \geq 0} A_3^{\nu}(y_1, x_2)\delta^{\nu}, \tag{6.96}$$

where

$$A_3^{\nu}(y_1, x_2) = \sum_{n \geq 0} a_{\nu}^{n}(y_1)\, u_3^{n}(y_1, x_2), \tag{6.97}$$

and $u_3^n(y_1, x_2)$ is the nth mode for a waveguide whose thickness is determined at y_1. This asymptotic construction is often called the JWKB technique. Note that the expansion has the form of a modulated group such as we encountered in (6.74). For simplicity, assume from now on that the guide is symmetric so that $h^{\pm} = h$ and $H^{\pm} = H$. Then the $u_3^n(x_2, y_1)$ are the cosine or sine functions first given in (6.3), with h replaced by $H(x_1)$ and γ_n replaced by $k\gamma_n$, $\gamma_n = n\pi/2H$. The boundary conditions at $x_2 = \pm H$ become

$$\mu\left[-\delta^2 \frac{dh}{dy_1}\frac{\partial u_3}{\partial y_1} \pm \frac{\partial u_3}{\partial x_2} \right] = 0. \tag{6.98}$$

Note that the boundary conditions are homogeneous, so that this equation is exact.

Now (6.96) is substituted into (6.93) and (6.98), and the coefficient of each power of δ set to zero. We are then led to the sequence of equations

$$\left[\mathcal{L}-\left(\frac{d\theta}{dy_1}\right)^2\right]A_3^\nu+i\left(\frac{d^2\theta}{dy_1^2}+2\frac{d\theta}{dy_1}\right)\frac{\partial A_3^{\nu-1}}{\partial y_1}+\frac{\partial^2 A_3^{\nu-2}}{\partial y_1^2}=0, \quad (6.99)$$

with their accompanying boundary conditions. We have introduced the operator $\mathcal{L}$, where $\mathcal{L}:=\partial^2/\partial x_2^2+1$, to make the structure of the equations clearer. The terms for which the superscripts are negative are zero. The equation corresponding to $\nu=0$ is

$$\mathcal{L}(A_3^0)=\beta_n^2 A_3^0, \quad (6.100)$$

with

$$\frac{\partial A_3^0}{\partial x_2}(y_1,\pm H)=0. \quad (6.101)$$

To arrive at (6.100) we have set

$$(d\theta/dy_1)^2=\beta_n^2. \quad (6.102)$$

This is an eikonal equation, similar in some respects to (2.43).

Equations (6.100) and (6.101) constitute an eigenvalue problem, identical to that solved in *Problem 6.1*. Note that β_n of (6.4) has become $k\beta_n$ and that now

$$\beta_n=\left[1-\gamma_n^2(y_1)\right]^{1/2}, \quad (6.103)$$

where $\gamma_n=n\pi/[2h_0+2h(y_1)]$. Further,

$$u_3^n(y_1,x_2)=N_n\begin{matrix}\cos\\ \sin\end{matrix}(\gamma_n x_2), \quad (6.104)$$

where N_n are constants, often determined by a normalization condition. The solution to (6.100) and (6.101) is then $A_3^0=a_0^n(y_1)u_3^n(y_1,x_2)$, with β_n^2 as the corresponding eigenvalue. Note that y_1 enters β_n through $h(y_1)$. The reader is asked to recall the earlier comments in *Problem 6.1* as to what the eigenvalue of interest is. The eikonal equation, (6.102), is integrated to give $\theta(y_1)$.

To find a_0^n we need to go to the next order in δ. This equation is

$$-\mathcal{L}(A_3^1)+\beta_n^2 A_3^1=i\frac{d^2\theta}{dy_1^2}A_3^0+2i\left(\frac{d\theta}{dy_1}\right)\left(\frac{\partial A_3^0}{\partial y_1}\right), \quad (6.105)$$

with boundary conditions, at $x_2 = \pm H$,

$$\pm \frac{\partial A_3^1}{\partial x_2} = i \frac{dh}{dy_1} \frac{d\theta}{dy_1} A_3^0. \tag{6.106}$$

We shall not seek terms of order higher than $\nu = 0$ here. Burridge and Weinberg (1977) indicate how higher-order terms are gotten.

Examining (6.105) and (6.106), we note that to have a bounded solution for A_3^1 it must not resonate with the A_3^0 forcing term. That is, it must be orthogonal in some sense to u_3^n. We use this fact to find a_0^n. First we introduce the inner product

$$[a, b] := \int_{H^-}^{H^+} a(x_2)\, b^*(x_2) dx_2 \tag{6.107}$$

and agree to normalize the eigenfunctions such that $[u_3^n, u_3^m] = \delta_{nm}$. The N_n are then given by

$$N_n = \{\epsilon_n[h_0 + h_1(y_1)]\}^{-1/2} \tag{6.108}$$

with $\epsilon_n = 2$ for $n = 0$, and $\epsilon_n = 1$, otherwise. Calculating the inner product $[\mathcal{L}(A_3^1), u_3^n]$ and using (6.105) and (6.106) gives the following transport equation for a_0^n, namely

$$\frac{d\theta}{dy_1} \frac{da_0^n}{dy_1} + \frac{1}{2} a_0^n \frac{d^2\theta}{dy_1^2} + a_0^n \frac{d\theta}{dy_1} \left\{ [\partial u_3^n / \partial y_1, u_3^n] + \frac{1}{2} \frac{dh}{dy_1} \left(|u_3^{n+}|^2 + |u_3^{n-}|^2 \right) \right\} = 0, \tag{6.109}$$

where the superscript plus and minus signs mean that these terms are evaluated at $x_2 = \pm H$, respectively. Note that

$$\left[u_3^n, \frac{\partial u_3^n}{\partial y_1} \right] = - \left[\frac{\partial u_3^n}{\partial y_1}, u_3^n \right] - \frac{dh}{dy_1} \left(|u_3^{n+}|^2 + |u_3^{n-}|^2 \right). \tag{6.110}$$

Taking the complex conjugate of (6.109) and adding the two gives the much simpler equation

$$\frac{1}{2} \frac{d}{dy_1} \ln \left(\frac{d\theta}{dy_1} \frac{d\theta^*}{dy_1} \right) + \frac{d}{dy_1} (a_0^n a_0^{n*}) = 0. \tag{6.111}$$

From this it follows immediately that

$$\left| \frac{d\theta}{dy_1} \right| |a_0^n|^2 = \text{constant}. \tag{6.112}$$

Equation (6.112) is a statement of energy conservation. In other words, to lowest order, no waves are reflected by the slowly changing width. The unknown constant would be determined from an initial condition at $x_1 = 0$.

To determine the argument θ_0^n of $a_0^n = |a_0^n| e^{i\theta_0^n}$, substitute this into (6.109). It is readily determined that θ_0^n is a constant. Thus

$$a_0^n = c_0^n e^{i\theta_0^n} / \beta_n^{1/2}, \tag{6.113}$$

where c_0^n is a real constant. Bringing the pieces together, we find the approximate expression for the nth mode is

$$u_3 \sim \exp\left[\frac{i}{\delta}\int^{y_1} \beta_n(s)\, ds\right] a_0^n(y_1) N_n(y_1) \begin{matrix}\cos\\ \sin\end{matrix} (\gamma_n x_2). \tag{6.114}$$

The β_n is given by (6.103), N_n by (6.108), and a_0^n by (6.113).

This particular approach can be extended to inplane elastic waves (Folguera and Harris, 1999). Among several newer features, arising when this is done is a reformulation of the inplane eigenvalue problem and the use of an inner product for the orthogonality condition that is not the norm of the space in which the problem is set.

6.6 Dispersion and Group Velocity

6.6.1 Causes of Dispersion

We first encountered dispersion in the study of a periodic structure and then later in the study of guided waves. The periodic structure could be imagined to be a rod possessing a periodic microstructure represented by the concentrated masses. It is the microstructure, in combination with the additional length scale introduced by it, that causes the dispersion. In fact, we could have examined the periodic structure by beginning with the equation

$$\partial_1^2 u_1 = \left[1 + \frac{M}{\rho}\sum_{-\infty}^{\infty} \delta(x_1 - nL)\right] \frac{1}{c_b^2} \partial_t^2 u_1. \tag{6.115}$$

In the end we should have solved it much as we did, but writing the equation in this form exhibits exactly how the microstructure enters the equation.

For guided waves the dispersion is caused by the geometrical or kinematical constraint that the waves must reflect from the boundaries in such a way that they reinforce one another to form a sustained wavefield. By considering only the nth waveguide mode for a closed waveguide, written as

$$u_{3n} = A_n \cos[(n\pi/h)x_2] P_n(x_1, t), \tag{6.116}$$

we find, by substituting this into the equation of motion, (6.1), that the propagating term $P_n(x_1, t)$ satisfies

$$c^{-2}\partial_t^2 P_n - \partial_1^2 P_n + (n\pi/h)^2 P_n = 0. \tag{6.117}$$

Note that (6.2) is automatically satisfied and that this equation captures the dispersive propagation.

The equation (6.117) is the Klein–Gordon equation, (6.27), and both (6.115) and (6.117) have the same form as do the Klein–Gordon equation and that for the flexural motion for a rod, (6.28). In *Problem 6.2* the reader showed that both these latter equations led to dispersive propagation.

Continuing, recall that in Section 6.1.2 we showed that the energy in the nth mode propagates at the group velocity C_n, and that in Section 6.3.2, using the method of stationary phase, we showed that the group velocity is that with which each ω_{ns} propagated to a point (x_1, t). While it would be misleading to assert that the energy in all wavefields or that the information encoded in them always propagates at the group velocity, this does occur for many wavefields. In this section, we study dispersion by focusing our attention on the meanings given the group velocity in the context of one dimensional, dispersive equations having a form such as (6.115) or (6.117).

Dispersive propagation is an extraordinarily rich subject. In addition to the work of Lighthill (1965) and Whitham (1974), which we introduce here, the reader is referred to Sommerfeld (1964a) and Brillouin (1960) for extended discussions of the dispersion of electromagnetic waves in dielectric materials.

6.6.2 The Propagation of Information

We consider a signaling problem. The wave, amplitude $u(x, t)$, begins at $x = 0$ as a square pulse modulating a harmonic carrier and propagates outward in the positive x direction. It is described at $x = 0$ by

$$u(0, t) = \Pi(t) \cos \omega_0 t, \tag{6.118}$$

with the modulation described by

$$\Pi(t) = H(t + T) - H(t - T). \tag{6.119}$$

As in previous instances, $H(t)$ is the Heaviside function. The dispersion relation has been found by seeking a solution in the form $Ae^{i(kx-\omega t)}$ and solving for $k = k(\omega)$. We construct a solution to the signaling problem as

$$u(x, t) = \frac{\Re}{\pi} \int_0^\infty {}^*u(0, \omega) e^{i[k(\omega)x - \omega t]}\, d\omega, \tag{6.120}$$

where the transform of $u(0, t)$ is given by

$${}^*u(0, \omega) = \frac{\sin[(\omega - \omega_0)T]}{\omega - \omega_0} + \frac{\sin[(\omega + \omega_0)T]}{\omega + \omega_0}. \tag{6.121}$$

Note that (1.36) has been used to write (6.120). If $\omega_0 T$ is large, ${}^*u(0, \omega)$ is concentrated within intervals near $\pm\omega_0$. In particular, the major contribution to the integral, (6.120), comes from $\omega \in (\omega_0 - \pi/T, \omega_0 + \pi/T)$. In this interval we can approximate the dispersion relation as

$$k(\omega) \approx k(\omega_0) + \frac{dk}{d\omega}(\omega_0)(\omega - \omega_0), \tag{6.122}$$

provided the derivative of $k(\omega)$ is well behaved and does not vanish. We can thus approximate (6.120) as

$$u(x, t) \approx \frac{\Re}{\pi}\left[e^{i(k_0 x - \omega_0 t)} M\left(t - \frac{x}{C}\right)\right], \tag{6.123}$$

where

$$M\left(t - \frac{x}{C}\right) \approx \int_{\omega_0 - \pi/T}^{\omega_0 + \pi/T} {}^*u(0, \omega) e^{-i(\omega - \omega_0)(t - x/C)}\, d\omega, \tag{6.124}$$

provided $\omega_0 T$ is large; $k_0 = k(\omega_0)$ and $C^{-1} = dk/d\omega(\omega_0)$. Using the change of variable $\Omega = \omega - \omega_0$, we can write (6.124) as

$$M(t - x/C) \approx \int_{-\pi/T}^{\pi/T} {}^*\Pi(\Omega) e^{-i\Omega(t - x/C)}\, d\Omega, \tag{6.125}$$

where ${}^*\Pi(\Omega)$ is the Fourier transform of $\Pi(t)$. Therefore, provided ${}^*u(0, \omega)$ is concentrated near ω_0, the modulation $\Pi(t) \approx M(t)$ propagates at the speed C, the group velocity.

This is essentially the kinematic argument put forward by Stokes (Sommerfeld, 1964b) to demonstrate that information propagates at the group velocity. There are shortcomings to this argument. The principal one is that ${}^*u(0, \omega)$ must be concentrated in the neighborhood of the carrier frequency ω_0, in order that there be a well-defined modulation. This, however, may not always be the case, and that raises the question as to whether or not the concept of group velocity has a more general significance.

6.6.3 The Propagation of Angular Frequencies

Whitham (1974) and Lighthill (1965) give a more general meaning to group velocity that is motivated by the stationary phase condition. However, they consider an initial value problem rather than a signaling problem so that their formulation is given in terms of a local wavenumber. *Problem 6.10* suggests why they take this approach. They propose both a kinematic theory describing the propagation of a local wavenumbers as well as a kinetic theory describing the evolution of a group. Little more than a dispersion relation and the assumption that far from from the source a wavefield evolves into a form described by $u(x, t) \sim \Re[A(x, t)e^{i\theta(x,t)}]$ is required.

Recall that following (6.74) we interpreted the group velocity as that with which a given local angular frequency, one that characterizes a group, propagates. We chose to describe the local angular frequency, rather than the local wavenumber, because we were working with a signaling problem and, therefore, it was natural to express the solution as a Fourier transform over the angular frequency ω. Nevertheless, the ideas of Whitham and Lighthill prove just as adept at describing the evolution of the local angular frequency and the group it characterizes.

A Signaling Problem

We again consider the signaling problem whose solution is given by (6.121). However, we now assume that $\omega_0 T$ is such that $^*u(0, \omega)$ is slowly varying. We approximate (6.120) for large x, assuming that t/x is fixed, using the stationary phase approximation (5.69). This gives

$$u(x, t) \sim \frac{\Re}{\pi}\left\{\frac{(2\pi)^{1/2}}{[x|d^2k/d\omega^2|]^{1/2}}\, {}^*u(0, \omega)e^{i[k(\omega)x-\omega t\pm\pi/4]}\right\}. \tag{6.126}$$

The stationary phase condition is $dk/d\omega = t/x$. The solution to this equation[3] gives $\omega = \omega(x, t)$, and $k = k(x, t)$ through the dispersion relation $k = k(\omega)$. We noted previously, following (6.73), that the stationary phase condition, written as

$$C(\omega) = x/t, \tag{6.127}$$

[3] The argument is really x/t rather than the more general (x, t). However, this latter notation will permit some generalizations in the subsection that follows.

is a mapping, for propagating waves, from the t plane to the ω plane for fixed x. In this calculation we emphasize this aspect of the stationary phase condition and have dropped the subscript s that was used previously to designate an ω satisfying this condition. The term $d^2k/d\omega^2$ is given by

$$\frac{d^2 k}{d\omega^2}(\omega) = -\frac{1}{C^2(\omega)}\frac{dC}{d\omega}(\omega). \tag{6.128}$$

We assume that $dC/d\omega > 0$ and take the minus sign in (6.126). If $dC/d\omega < 0$, then the reasoning remains the same, though signs change here and there. Moreover, we are, throughout, assuming that $dC/d\omega \neq 0$. Thus, $u(x,t) \sim \Re[A(x,t)e^{i\theta(x,t)}]$, where

$$A(x,t) = \frac{1}{\pi}\left\{\frac{(2\pi)^{1/2}}{[x|d^2k/d\omega^2|]^{1/2}}\right\}^* u(0,\omega)e^{-i\pi/4} \tag{6.129}$$

and

$$\theta(x,t) = k(\omega)x - \omega(x,t)t. \tag{6.130}$$

A Kinematic Theory

Assuming that we know $\theta(x,t)$ and a dispersion relation $k = k(\omega)$, we can define a local angular frequency and a local wavenumber as

$$\omega(x,t) = -\partial_t\theta, \qquad k(x,t) = \partial_x\theta. \tag{6.131}$$

If these definitions are to be consistent with one another,

$$\partial_t k + \partial_x \omega = 0, \tag{6.132}$$

or, by using the dispersion relation,

$$\partial_t \omega + C(\omega)\partial_x \omega = 0, \tag{6.133}$$

where $C = d\omega/dk$. This is a partial differential equation for the angular frequency ω. Note that (6.132) has the form, defined by (6.30), of a conservation equation, and that (6.133) captures the idea suggested by the stationary phase condition that the frequency ω propagates at the group velocity C.

Equation (6.133) is solved by the method of characteristics (Whitham, 1974; Zauderer 1983). Provided the characteristics do not intersect, the solution can be gotten quite simply.

1. Assume that there exists a curve $\mathcal{S}$ in the (x,t) plane such that along $\mathcal{S}$, $dt/dx = C^{-1}$. Then (6.133) becomes $d\omega/dt = 0$ along this curve.

2. $\omega(x, t)$ is therefore constant along $\mathcal{S}$ and, because $\mathcal{C}$ is constant if ω is, $\mathcal{S}$ is given by $t = x\mathcal{C}^{-1}(\omega) + t_0$, where t_0 is the intercept with the t axis. Hereafter we write $\mathcal{S}(t_0)$ to identify that member of the family of curves with intercept t_0.
3. Assume that an initial distribution of frequencies $\omega(0, t_0) = f(t_0)$ is given. Along each $\mathcal{S}(t_0)$, $\omega(0, t_0) = \omega(x, t)$. It follows then that $\omega(x, t) = f[t - x\mathcal{C}^{-1}(\omega)]$ along $\mathcal{S}(t_0)$. Variation of t_0 gives a solution throughout (x, t).

It is not an explicit solution for ω. Nevertheless, it indicates its evolution. Lastly, if we assume that $f(t_0)$ is localized near the origin so that $t_0/x \ll 1$ for x large and t/x fixed, we recover the stationary phase condition, (6.127).

We can also define a local phase velocity. Assume $\theta(x, t) = \theta_0$, a constant. Implicit differentiation gives

$$c(x, t) = -\partial_t\theta/\partial_x\theta, \tag{6.134}$$

where the phase velocity $c = dx/dt$. Thus $c = \omega/k$ where ω and k are the local frequency and wavenumber defined by (6.126).

We call lines of constant angular frequency *group lines* and those of constant phase *phase lines. Reverting to the notation of* Section 6.3.2, we again consider the stationary phase approximation to the transiently excited waves in a closed waveguide. The local angular frequency determined by the stationary phase condition is given by (6.129). That expression, rewritten here, is

$$\omega_n(x_1, t) = \frac{(cn\pi/h)}{[1 - (x_1/ct)^2]^{1/2}}. \tag{6.135}$$

The phase, given by the argument of the cosine in (6.72), is

$$\theta_n(x_1, t) = -\frac{x_1 n\pi}{h}\left[\left(\frac{ct}{x_1}\right)^2 - 1\right]^{1/2}. \tag{6.136}$$

For both expressions $\omega_n \in (cn\pi/h, \infty)$. The group lines or curves of constant local angular frequency $\omega_n = \omega_{n0}$ are given by

$$\frac{x_1}{ct} = \left[1 - \left(\frac{cn\pi}{\omega_{n0}h}\right)^2\right]^{1/2}. \tag{6.137}$$

The curves of constant phase $\theta_n = \theta_{n0}$ are given by

$$c^2t^2 - x_1^2 = \left(\frac{\theta_{n0}h}{n\pi}\right)^2. \tag{6.138}$$

These two families of curves are quite different so that, once again, we note that a phase and a group evolve differently in a dispersive environment.

A Kinetic Theory

Previously, we explicitly showed that for a closed waveguide the averaged energy of a particular mode propagates with the group velocity for that mode, (6.25). Auld (1990) uses a reciprocity relation to prove this result in somewhat greater generality. Lighthill (1965) gives a general argument showing that, for any wave structure or environment that can be characterized by a Lagrangian and for which equipartition of energy occurs, the group velocity is the velocity with which the average energy propagates. Aki and Richards (1980) give a restricted version of this argument that is applicable to waveguide modes. Here we continue to follow Whitham (1974) and Lighthill (1965) and advance an argument using the stationary phase condition.

We define a quantity $\mathcal{Q}$ as follows.

$$\mathcal{Q}(x) = \int_{t_1}^{t_2} A(x, t)A^*(x, t)\omega^2(x, t)dt, \tag{6.139}$$

where $t_2 > t_1$. This is reminiscent of an energy term. For a time harmonic, plane wave and wave system for which equipartition of energy occurs, this term is proportional to the time-averaged energy density. Recalling our earlier comment that (6.127) is a mapping from t to ω, we change the variable of integration in (6.138) to ω. Noting that

$$dt = -\frac{x}{C^2(\omega)}\frac{dC}{d\omega}(\omega)d\omega, \tag{6.140}$$

we see that (6.139) becomes

$$\mathcal{Q}(x) = \frac{2}{\pi}\int_{\omega_2}^{\omega_1} |^*u(0, \omega)|^2\omega^2 d\omega \tag{6.141}$$

For $dC/d\omega > 0$, $\omega_1 > \omega_2$. This is a remarkable result. It indicates that for a signal, such as a fragment of speech, the energy contained within a given frequency band is constant when the frequency band remains between the two group lines identified by $t_i = xC^{-1}(\omega_i)$, where $i = 1, 2$. This then is a generalization of the idea that time-averaged energy propagates at the group velocity.

While the kinematic and kinetic descriptions of group velocity put forward by Whitham (1974) and Lighthill (1965) appear quite complete, they assume that attenuation is not present. When dispersion arises from material properties, it is very often accompanied by frequency-dependent attenuation. The same physical mechanisms are responsible for both dispersion and attenuation so that they cannot be easily disentangled, unless the attenuation is quite weak. The propagation of linearly viscoelastic waves (Hudson, 1980) is an example that indicates the tangling of these two effects. In such cases, attributing a meaning to group velocity can become much more difficult. An ab initio calculation may be the only guide (see also footnote 2).

Problem 6.10 Dispersion 2

As we indicated previously, Whitham (1974) and Lighthill (1965) use an initial value problem to advance their ideas about group velocity. Thus they write that the wavenumber propagates at the group velocity and that the energy confined between two group lines defined by two fixed k_i is constant. The following two problems indicate why they develop their ideas in this way.

Problem 1. Consider once again the Klein–Gordon equation, (6.27). Solve the initial value problem for this equation, given the initial conditions

$$\varphi(x,0) = \delta(x), \qquad \partial_t \varphi(x,0) = 0. \tag{6.142}$$

For $t < 0^-$, φ and $\partial_t \varphi$ are identically zero. Thus show that the appropriate solution is

$$\varphi = \frac{1}{\pi} \int_0^\infty \cos(kx)\cos[\omega(k)t]\,dk, \tag{6.143}$$

where $\omega = \omega(k)$ is the dispersion relation. Note that (6.143) is to be interpreted as a generalized function or distribution. Note that k, and not ω, is the transform variable.

Using the method stationary phase, show that this solution for $t \to \infty$, with x/t positive and fixed, can be put into the form

$$\varphi \sim \Re\left[A(x,t)e^{i\theta(x,t)}\right]. \tag{6.144}$$

Give explicit expressions for $A(x,t)$ and $\theta(x,t)$.

Problem 2. Repeat *Problem 1* for the equation for flexural motion in a rod, (6.28).

References

Aki, K. and Richards, P.G. 1980. *Quantitative Seismology, Theory and Methods*, Vol. 1, pp. 286–292. San Francisco: Freeman.

Auld, B.A. 1990. *Acoustic Fields and Waves in Solids*, 2nd ed., Vol. 2, pp. 203–206. Malabar, FL: Krieger.

Brekhovskikh, L.M. 1980. *Waves in Layered Media*, 2nd ed., pp. 299–325. New York: Academic.

Brillouin, L. 1960. *Wave Propagation and Group Velocity*, 2nd ed. New York: Academic.

Burridge, R. and Weinberg, H. 1977. Horizontal rays and vertical modes. In *Wave Propagation and Underwater Acoustics*, pp. 86–152, ed. J.B. Keller and J.S. Papadakis. New York: Springer.

Dai, H.-H. and Wong, R. 1994. A uniform asymptotic expansion for the shear-wave front in a layer. *Wave Motion* **19**: 293–308.

deSanto, J.A. 1992. *Scalar Wave Theory*, pp. 162–179. New York: Springer.

Folguera, A. and Harris, J.G. 1999. Coupled Rayleigh surface waves in a slowly varying elastic waveguide. *Proc. R. Soc. Lond.*, A **455**: 917–931.

Friedman, B. 1956. *Principles and Techniques of Applied Mathematics*. New York: Wiley.

Hudson, J.A. 1980. *The Excitation and Propagation of Elastic Waves*, pp. 188–222. Cambridge: University Press.

Lighthill, M.J. 1965. Group velocity. *J. Inst. Maths. Appl.*, **1**: 1–28.

Miklowitz, J. 1978. *The Theory of Elastic Waves and Waveguides*, pp. 178–200 and 409–466. New York: North-Holland.

Sommerfeld, A. 1964a. *Optics, Lectures on Theoretical Physics*, Vol. IV, pp. 273–289. Translated by O. LaPorte and P.A. Moldauer. New York: Academic.

Sommerfeld, A. 1964b. *Mechanics of Deformable Bodies, Lectures on Theoretical Physics*, Vol. II, pp. 184–191. Translated by G. Keurti. New York: Academic.

Tolstoy, I. and Clay, C.S. 1987. *Ocean Acoustics, Theory and Experiment in Underwater Sound*, pp. 76–80 and elsewhere. New York: American Institute of Physics.

Whitham, G.B. 1974. *Linear and Nonlinear Waves*, pp. 363–402. New York: Wiley-Interscience.

Zauderer, E. 1983. *Partial Differential Equations of Applied Mathematics*, pp. 35–77. New York: Wiley-Interscience.

Index

κ, defined, 40
κ_r, defined, 55

Abelian theorem, 82, 90, 107
allied function, 46
angular frequency
  defined, 9
angular-spectrum; *see* plane-wave, 24
asymptotic approximation of integrals, 86–96
  end point contribution, 95
  Fresnel integral, 119
  integration by parts, 87
  stationary phase, 94–95
    contour, 92
    waveguide modes, 136
  steepest descents, 90–94
    branch cut contribution, 99
    contour, 91, 98
    pole contribution, 99
  Stokes' phenomenon, 119
  uniform, 108
  Watson's lemma, 82, 87–90
  wavefront approximation, 82, 95
asymptotic power series, 30
asymptotic ray expansion, 28–34, 112
  compressional, 29
  shear, 33
average
  for a closed waveguide mode, 126
  Lagrangian density, 126
  time average for a plane wave, 23

boundary layer; *see* matched asymptotic expansion, 109
branch cuts, defining, 52–55
buried harmonic line of compression, 82–86, 96–101
  asymptotic approximation of the scattered compressional wave, 98
  asymptotic approximation of the scattered shear wave, 99

Cagniard–deHoop technique, 77–82
  contour, 81
  inversion, 79–82
caustic, 31, 85, 100
center of compression
  three-dimensional, 61
  two-dimensional, 62
complex unit vector, 23
compressional wave, defined, 5
critical angle
  incidence, 43
  reflection, 44
  refraction, 43
cut-off frequency, 125, 133
cylindrical wave, 34

diffraction at an edge, 101–119
  diffracted wave, 108
  diffraction coefficient, 110
  diffraction integral, 108; *see* Fresnel integral
  geometrical theory, 110
dispersion; *see* velocity, group, 13
  and stationary phase, 136
  anisotropic medium, 128
  anomalous, 138
  causes of, 150
  closed waveguide, 122, 125
  from the poles of an integral representation, 144
  geometrical, 150
  microstructure, 150

dispersion (*cont.*)
open waveguide, 130, 131
periodic structure, 18

eigenvalue problem, 122
eigenfunction
continuous expansion, 139
discrete or mode expansion, 134
eigenvalues, 123
discrete and continuous, 145
eigenvalues and eigenfunctions
discrete, 122
discrete and continuous, 131
eikonal equation, 30, 148
energy relations, 5
kinetic energy density, defined, 6
averaged for a closed waveguide mode, 126
averaged for a plane wave, 23
conservation law, 6, 127
energy flux during critical refraction, 45
energy flux during reflection, 41
energy flux, defined, 6
energy in a band of frequencies, 156
equipartition of energy, 126
for a transient plane wave, 22
for the flexural motion equation, 128
for the Klein–Gordon equation, 128
internal energy density, defined, 6
equations of motion, 1–6
boundary conditions, 2
dilitational motion, 2
one-dimensional, 3
rotational motion, 2
two-dimensional, 4
antiplane motion, 4
inplane motion, 4
extinction theorem, 75

farfield
compact source, 61, 65
of an edge, 114
Fermat's principle, 101
flexural motion equation, 48, 127, 157
Fourier transform
space, 10
three-dimensional, 58
time, 7
frequency
defined, 9
local, 155
Fresnel integral, 109, 116–119

gamma function, 87
Gaussian beam; *see* plane-wave representations, 25
geometrical theory of diffraction; *see* diffraction at an edge, 111

Green's function, 60, 62
method of images, 68
Green's tensor, 58, 62
correct and incorrect, 66
elastic fluid, 74
stress, 60
three-dimensional, 58
guided waves; *see* waveguide, 121

Helmholtz theorem, 5

initial value problem, 157
inner expansion; *see* matched asymptotic expansion, 71
integral equations, 67, 76
integral representation
scattering problem, 65–67
source problem, 64–65

JWKB asymptotic expansion
eigenvalue problem, 148
rays and modes, 147

Klein–Gordon equation, 127, 150, 157

Lagrangian density, 126
Lamb's problem; *see* buried harmonic line of compression, 82
Laplace transform
time, 7
two-sided, 78
leaky wave; *see* waveguide, open, 146
limiting absorption, principle of, 62–63, 65, 135
and the Wiener–Hopf method, 104–108
outgoing wave, 58, 59, 79
line of compression; *see* center of compression, 83
longitudinal wave; *see* compressional, 22
Love wave; *see* waveguide, open, 131

matched asymptotic expansion, 112–116
boundary layer, 109, 111, 113
inner, outer coordinates, 71, 114
inner, outer expansion, 71, 114
matching, 115
Schrodinger equation, 114
stretching transformation, 114

nearfield of an edge, 72, 104, 110, 113

outer expansion; *see* matched asymptotic expansion, 71
outgoing wave; *see* limiting absorption, 59

partial waves, 17, 123, 140, 145
periodic structure, 15–18
effective wavenumber, 18

phase matching, 39–40
phase, its meaning, 29
plane wave
  homogeneous, 22
  inhomogeneous or evanescent, 22, 43, 96
  time-dependent, 20
  time-harmonic, 22
plane-wave representations, 24–28
  cylindrical wave, 28, 140
  Gaussian beam, 24
  in an open waveguide, 142, 146
  spherical wave, 27, 28
  waves scattered from a surface, 84
Poisson summation formula, 11
polarization
  change in a ray expansion, 33
  defined, 20
potentials, displacement, 5
  scalar potential, 5
  vector potential, 5
propagation matrix, 16

radiation conditions, 65; *see* limiting absorption, 65
ray; *see* asymptotic ray expansion, 30
  defined, 30
  fan of diffracted rays, 110
  ray tube, 33
  rays and modes, 146
Rayleigh wave, 48–52, 96, 101
  Rayleigh function, 51
  Rayleigh equation, 49
  time-harmonic, 49
  transient, 50
  two-sided wave, 51
reciprocity, 57
reflection
  antiplane shear coefficient, 44
  antiplane shear wave, 42
  compressional coefficient, 40
  compressional wave, 38
  from a layer, 133
  from a plate in a fluid, 47
  inplane shear coefficient, 40
  inplane shear wave, 38
refraction
  antiplane shear coefficient, 44
  antiplane shear wave, 42
  interfacial wave, 45
  two-sided wave, 46

scalar potential; *see* potentials, displacement, 5
scattering
  Bragg, 17
  from a lumped mass, 13
  from a strip, 67
  from an elastic inclusion, 72
  scattering matrix, 14
shear wave, defined, 5
signaling problem, 151–157
  defined, 24
slowness
  defined, 20
  surface, 39, 44
  vector, 20, 39
Sommerfeld transformation, 27, 59, 84, 96, 142
spherical wave, 34
standing wave, defined, 9
stationary phase; *see* asymptotic approximation of integrals, 94
steepest descents; *see* asymptotic approximation of integrals, 90
strain tensor, defined, 2
stress tensor, defined, 1
surface wave; *see* Rayleigh wave, 51
  guided by an impedance boundary, 51

Tauberian theorem; *see* Abelian theorem, 82
traction, defined, 1
transforms, Fourier and Laplace; *see* Fourier transform, Laplace transform, 6
  defined, 6–10
transmission coefficient; *see* refraction, 44
transmission matrix, 15
transport equation, 30, 149
transverse resonance principle, 124
transverse wave; *see* shear, 22

uniqueness in an unbounded region, 63, 68–72
  edge conditions, 69, 104, 107
  no edges, 68

vector potential; *see* potentials, displacement, 5
velocity
  energy transport, 22, 24, 125
  fastest, 137
  group; *see* dispersion, 18
    and stationary phase, 137
    group lines, 155
    kinematic theory, 154
    kinetic theory, 156
    periodic structure, 18
    propagation of angular frequencies, 153–157
    propagation of information, 151
    propagation of wavenumbers, 153
    waveguide mode, 125
  phase
    local, 155
    phase lines, 155
    waveguide mode, 125
vibration, not a wave, 11

Watson's lemma; *see* asymptotic approximation of integrals, 87
wave kinematics, defined, 20
wavefront
  curvature, 32
  defined, 30
  plane, spherical, etc., 20
waveguide
  closed, 121
    evanescent mode, 135
    excitation, 134–139
    harmonic excitation, 134
    mode expansion, 134
    transient excitation, 135
  laterally inhomogeneous, 146
  mode, 122
  open, 129
    excitation, 139–146
    leaky modes, 146
    modal representation, 144
    plane-wave representation, 142, 146
    ray representation, 141, 145
wavelength, defined, 9
wavenumber
  complex, 22
  defined, 8
  effective, periodic structure, 18
  lateral, 122
  local, 155
  transverse, 122
wavevector
  defined, 22
  real and complex, 22
Wiener–Hopf method, 104–108

Printed in the United States
By Bookmasters